Science and Fiction

Science and Fiction – A Springer Series

This collection of entertaining and thought-provoking books will appeal equally to science buffs, scientists and science-fiction fans. It was born out of the recognition that scientific discovery and the creation of plausible fictional scenarios are often two sides of the same coin. Each relies on an understanding of the way the world works, coupled with the imaginative ability to invent new or alternative explanations—and even other worlds. Authored by practicing scientists as well as writers of hard science fiction, these books explore and exploit the borderlands between accepted science and its fictional counterpart. Uncovering mutual influences, promoting fruitful interaction, narrating and analyzing fictional scenarios, together they serve as a reaction vessel for inspired new ideas in science, technology, and beyond.

Whether fiction, fact, or forever undecidable: the Springer Series "Science and Fiction" intends to go where no one has gone before!

Its largely non-technical books take several different approaches. Journey with their authors as they

- Indulge in science speculation—describing intriguing, plausible yet unproven ideas;
- Exploit science fiction for educational purposes and as a means of promoting critical thinking;
- Explore the interplay of science and science fiction—throughout the history of the genre and looking ahead;
- Delve into related topics including, but not limited to: science as a creative process, the limits of science, interplay of literature and knowledge;
- Tell fictional short stories built around well-defined scientific ideas, with a supplement summarizing the science underlying the plot.

Readers can look forward to a broad range of topics, as intriguing as they are important. Here just a few by way of illustration:

- Time travel, superluminal travel, wormholes, teleportation
- Extraterrestrial intelligence and alien civilizations
- Artificial intelligence, planetary brains, the universe as a computer, simulated worlds
- Non-anthropocentric viewpoints
- Synthetic biology, genetic engineering, developing nanotechnologies
- Eco/infrastructure/meteorite-impact disaster scenarios
- Future scenarios, transhumanism, posthumanism, intelligence explosion
- Virtual worlds, cyberspace dramas
- Consciousness and mind manipulation

More information about this series at http://www.springer.com/series/11657

Zoran Živković

First Contact and Time Travel

Selected Essays and Short Stories

 Springer

Zoran Živković
Belgrade, Serbia

ISSN 2197-1188   ISSN 2197-1196   (electronic)
Science and Fiction
ISBN 978-3-319-90550-1   ISBN 978-3-319-90551-8   (eBook)
https://doi.org/10.1007/978-3-319-90551-8

Library of Congress Control Number: 2018943298

Cover photo: By courtesy of Nuno Ferreira Santos (Lisbon 2016)

Printed on acid-free paper

This Springer imprint is published by the registered company Springer International Publishing AG part of Springer Nature.
The registered company address is: Gewerbestrasse 11, 6330 Cham, Switzerland

To Dragoljub Kojčić, my dear friend

Preface

The two main parts of this book—essays and fiction—originated during two rather distant periods of my life. With one exception, all the nonfiction pieces were written in the second half of the 1970s, nearly twenty years before I embarked on fiction. At that time, in my late twenties and early thirties, I was a young scholar working on my MA and PhD theses. I hadn't even remotely considered the possibility of becoming an author myself.

Strange as it might seem today, my area of academic interest was then revolutionary: science fiction. Although by that time the SF genre had already abandoned its origins in pulp literature and started to produce works of indisputable artistic value, it was still far from being a favorite subject in proverbially conservative academic circles.

I was very fortunate indeed to have an exceptional mentor, professor Nikola Milošević, who, although by no means an expert in science fiction himself, realized that it possessed the potential to offer new insights into some of the fundamental dilemmas, not only of the art of prose, but also, more generally, in his principal area of interest—the history of ideas.

In my PhD thesis ("The Origin of Science Fiction as a Genre of Artistic Prose," 1982) I tried to explain a unique phenomenon—how of all genres of pulp literature only science fiction had succeeded in becoming art. In the long run, however, my MA thesis had the quality of a genuinely pioneering study: "Anthropomorphism and the First Contact Theme in the SF Works of Arthur C. Clarke," 1979. Sir Arthur told me in one of his letters that, to the best of his knowledge, this was also the first academic paper ever written on his SF works. (Although flattered, I never cared to check because I don't feel that precedence is really very important in these matters.)

Apart from first contact, I was also interested in a second theme unique to science fiction—time travel (or, to use Lem's beautiful neologism, chronomotion). In my last, brief essay on science fiction (1995) I recapitulated all the subaspects of this very challenging theme in order to identify those that might have greater literary potential.

For a decade and a half (1975–1990) I tried my hand at every aspect of science fiction—but one. I wrote several books on it including a two-volume set: *The Encyclopedia of Science Fiction*. I translated more than 40 SF books, I was a critic, a reviewer, and a commentator on the SF genre, I hosted a TV series on the history of SF cinema and attended numerous conventions, conferences, festivals, and so on.

But I was never a science fiction writer.

A somewhat simplified answer to the inevitable question as to how I could possibly not become an SF writer with such a background is that by the time I began to write my first piece of fiction in 1993, science fiction had already gone into decline. This is not the place to elaborate on this, but it is my view that science fiction no longer exists. It belongs to the history of literature as one of the two great movements of the art of "fantastika" in the twentieth century. (The other is, of course, magical realism.) In the twenty-first century, we don't write science fiction because we don't need it. We live it. It is all around us. For better or worse.

In any case, what I write is not science fiction. (Curiously enough, no matter how often I repeat this simple fact, for the great majority of my compatriots who care to have an opinion I will forever remain a science fiction writer. Particularly for those who, for one reason or another, have had neither the opportunity nor the interest to read any of my 22 works of fiction.) I consider myself a writer without prefixes. Simply a writer.

On the other hand, not being an SF writer doesn't mean that I avoid themes introduced by science fiction. On the contrary, it is precisely through its new approaches to old SF themes that the new "fantastika" of the twenty-first century, which still doesn't even have a name, is slowly but surely taking its final shape.

If I had been an SF writer, I would never have been able to write *Time Gifts* or "The Puzzle"—my variations on the two pivotal science fiction themes: time travel and first contact. It took a long time to complete what I started back in the 1970s as an essayist. But completion would never have been possible without my being a writer.

Novi Beograd, Serbia   Zoran Živković
March 2018

Contents

Part II   Fiction

Part I

Essays

First Contact

Let us therefore tell the truth to ourselves: we are not searching for "all possible civilizations," but above all those which are anthropomorphic. We introduce the law and order of experiment into Nature and after phenomena of this kind we want to meet beings similar to ourselves. Nevertheless, we do not succeed in perceiving them. Do they in fact exist at all? There is indeed something deeply saddening in the silence of the stars as an answer to that question, a silence which is so complete as to be eternal.

—Stanislaw Lem, *Summa Technologiae*

Sometimes, in the dark of the night, I lie awake and wonder if different intelligences can communicate at all; or, if I've had a particularly bad day, whether the phrase 'different intelligences' has meaning at all.

—Isaac Asimov, *Gods Themselves*

1

The Theme of First Contact in the SF Works of Arthur C. Clarke

Sooner or later, it was bound to happen.

Arthur C. Clarke, *Rendezvous with Rama*

1.1 Introduction

The "first contact" theme in science fiction is characterized by its two generically different kinds of protagonist: the human and the alien. The notion of alien characters in fiction introduces a fundamental confusion, the resolution of which depends on what we would term the "artistic coherence" of the "first contact" theme: namely, is it at all possible to imagine and conjure up from a human perspective something essentially alien? The degree of difference between the human and alien protagonists in the "first contact" does not have to be absolute, of course, but the problem then changes in the quantitative and not the qualitative sense.

The human/nonhuman confusion appears on two levels, that is, in the context of the two different viewpoints attributing human characteristics to the alien which can exist in a work of sf. One is the perspective of the human characters in the work, and the other is of the author himself, as present in the narrative voice. From each of these perspectives, aliens can be ascribed human

"The Theme of First Contact in the SF Works of Arthur C. Clarke." Written in 1978–79. Originally published in Serbian in 1985 in *Prvi kontakt/First Contact*, Književne novine, Belgrade, Serbia. First published in English in "The New York Review of Science Fiction", New York, USA, in two parts: February 2001, 8–13, and March 2001, 10–17.

© Springer International Publishing AG, part of Springer Nature 2018
Z. Živković, *First Contact and Time Travel*, Science and Fiction,
https://doi.org/10.1007/978-3-319-90551-8_1

characteristics, but these two anthropomorphizations will not have an identical effect on the coherence of the first contact theme.

The whole skill of writing sf works with a "first contact" theme is in fact embodied in avoiding the anthropomorphic pitfalls which appear during the process of imagining and conjuring up alien characters with independent status. Furthermore, of course, the question arises as to uttermost limits, and whether it is at all possible to portray a truly alien entity by literary means.

When the human characters anthropomorphize the alien characters, the "first contact" theme serves as a means of artistic expression, in the sense that this factor is used as the best possible motivation for certain human characteristics and states. If, however, the anthropomorphization is from the perspective of the narrative voice, the coherence of the first contact theme is often disturbed, inasmuch as it rests on the fundamental assumption of alienness of the nonhuman protagonists.

There does exist, however, a kind of anthropomorphization of an alien entity from the perspective of the narrative voice that does not imperil the coherence of a work. This appears in those works in which the author uses the alien as a mirror, and in which the nonhuman character does not have an independent status but exists only because, through its mediation, one can make a statement about people. When, in contrast, the alien does have independent status, or when its role does not consist of the mere illustration of something basically human, then it is only in this case that one can speak of the real meaning of "first contact."

One of the authors who most thoroughly examines this confusion in his first contact stories is Arthur C. Clarke. Probably his most successful work in this respect is his famous novella "A Meeting with Medusa." To show to what extent Clarke had previously avoided anthropomorphic difficulties, we will first consider some of his short stories of a reflective type which focus on revealing basic aspects of the emergence of these factors in human consciousness.

With regard to the nature of man's relations towards an alien entity, one can differentiate three kinds of anthropomorphism in three Clarke parables of "first contact": anthropocentrism, anthropochauvinism, and simple anthropomorphism.

The first type, anthropocentrism, regards human beings as the central fact and final aim of the universe and so is a priori hostile towards the possibility of the existence of any other forms of intelligent life. The second type, anthropochauvinism, does not exclude this possibility but assumes the superior position of man in relation to any alien being. Finally, in the context of the third type, anthropomorphism, the possibility is allowed not only of the existence of alien entities, but also of their superiority in relation to man. Any possible intellectual intuition about aliens is, however, thwarted by innate deficiencies in the anthropomorphic nature of man's cognitive apparatus, as all aliens are seen in terms of human cognition. As examples of the types of anthropomorphic deficiencies, we will discuss three stories by Clarke: "Report on Planet Three," "Crusade," and "History Lesson."

1.2   Three Short Stories

1.2.1 "Report on Planet Three"

In "Report on Planet Three" there are two narrative perspectives. The first is represented by a document written by a certain Martian scientist at a time when our own civilization was still in its infancy, devoted to a consideration of the possibility of the existence of life in the third planet of the Solar System. The second perspective is that of the translator from Earth through his comments on the document, which was found in the ruins of the now-destroyed Martian civilization.

Although only the translator is aware of the "encounter" of two cosmic civilizations, the story focuses on the report of the scientist from Mars. The report represents a conspicuous example of orthodox planetary provincialism, the special feature of which is that it is expressed exactly from the standpoint of "official science," which has in this case already reached a level where it has mastered the technique of interplanetary flight.

The geophysical data on Earth, upon which the Martian bases his consideration of the possibility of life on Planet Three, have been obtained by valid astronomical methods. Troubles arise, however, when he gets down to interpreting these data—an interpretation in which the weak points are easily perceptible, as they are founded on inappropriate criteria.

The fallacy is reflected in the criteria for evaluating the conditions for possible life on Earth. The Martian scientist is conditionally in the right when he asserts that life will never develop on the Solar System's third planet—because what he has in mind by "life" is a notion valid exclusively in the biophysical context of Mars. To give it more general meaning outside this context points directly to the existence of certain deficiencies of interpretation.

The form of life native to the "red planet" cannot indeed develop on Earth, but this does not mean that it is unable in any way to nurture some other forms of life. The presence of water, oxygen and the hot regions round the Equator—those things chosen by the Martian scientist as his strongest arguments—not only did not prevent the beginning of life on our planet, but in fact represent the essential conditions for its birth and development. It is precisely in these comparisons that the provincial criteria of the document entitled Report on Planet Three suffer a total collapse: when conditions for the birth of life are in question, Mars has already been shown as unsuitable in principle to be a yardstick for Earth.

What, however, lies deeper within Clarke's story and makes it a good example for our consideration? What is the real cause of this Mars-centric fallacy? Is it, simply, a matter of intellectual immaturity and incapacity to outgrow the local circumstances of one's own world which, in an inappropriately provincial way, proclaim themselves as a yardstick of the whole universe, or is there possibly something else involved?

That the "errors" of the main character of the story "Report on Planet Three" are also influenced by other factors, which can't be reduced to mere intellectual limitation, is demonstrated by certain features of his report. The first part of the document, in which the Martian scientist merely cites the geophysical characteristics of our planet, sticking to the factual plane during this process, already reveals a hostile attitude towards the existence of life on Earth. The uncompromising negativity appears predominantly in the intonation and method of reporting the data. But this does not diminish its effect.

For example, when he needs to describe the particular colors of our planet, the scientist from Mars uses rather vague terms which, so the translator from Earth asserts, can be translated alternatively as "hideous" and "virulent." The entire further series of data—the existence of a large quantity of water on the Earth's surface, the density of the atmosphere, the presence of "poisonous and very reactive" oxygen, the "intolerable temperatures" at the Equator, and the "gigantic" force of gravity—are worded in such a way as to suggest a picture of Earth as a special kind of hell.

The irony in the report reaches its peak in a request for "scientific objectivity." "However, let us be open-minded"—says the author of the Report on Planet Three—"and prepared to accept even the most unlikely possibilities, as long as they do not conflict with scientific laws." "Scientific objectivity," which ought to be a valid criterion for a degree of "open-mindedness," is a calculated alibi for the lowest form of xenophobic provincialism, which is expressed when he begins to consider the hypotheses on the possibilities for the existence of higher intelligent forms on Earth, as a specific counterpart to the Martians.

The very calculated devaluation of these ideas is reflected in the fact that, without exception, they are ascribed to the authors of science fiction and speculative works, the worth of which has already been determined by the very fact that they appear as an open counterweight to "official science," which the Martian scientist refers to abundantly and on any occasion. The real nature of his fallacy becomes clear exactly on this plane. There is no question of any intellectual limitation but an attitude which does not flinch from "overlooking" the facts, simply in order to preserve an illusory adherence to one particular genocentric picture of the world.

The thing, however, which to a certain degree remains unclear within such an interpretation of the work is the overstressed anthropomorphization, as much of the Martian scientist as of his document Report on Planet Three, and of the broader framework which this document assumes. There is only one satisfactory answer to this illusory inexplicability: The story in fact represents a parable of man at the beginning of the cosmic era, and the provincial nature of the document Report on Planet Three displays all the features of orthodox anthropomorphism.

This exchange of roles was used by Clarke because taking the example of Earth as a foreign planet reveals contradictions that arise when local yardsticks are unreservedly proclaimed to be universal. Only when one realizes that it is in fact humanity's perspective which is involved in "Report on Planet Three" does the other, more hidden system of motivation for the lowest aspect of anthropomorphism become evident.

In addition to human intellectual limitations, which at least in principle do not have to be unbridgeable obstacles, Clarke introduces one more element with a different nature and effect: This is man's need to defend at any cost his dominant position in the natural order, a position seriously imperiled by the appearance of some new intelligent entity.

Human ambition expresses itself through intolerance and open disregard for anything that would directly or indirectly cast into doubt his status as the only intelligent being. This is thus the most orthodox and lowest form of anthropomorphism—anthropocentrism.

1.2.2 "Crusade"

We encounter a more complex form of anthropomorphism which no longer takes an a priori hostile attitude towards other kinds of intelligent life, but still retains the idea of superiority, an idea in this case based on a conviction about an exclusively "natural" origin, in the story "Crusade."

The protagonist in this work, a gigantic entity of electronic intelligence, has evolved in a world that is a natural "computer's paradise." This cosmic body is situated far away from the red-hot centers of the galaxies and the temperature on it reaches only a fraction of a degree above absolute zero. The superconductivity that prevails in its seas of liquid helium has created the perfect environment for the birth of mechanical intelligence. This is a special kind of "natural computer," capable of the faultless execution of gigantic analytical operations.

The enormous analytical potential of this computer predominates in its being to an extent which excludes "personal identity" and the capability of an emotional disposition towards the world. The conclusions that this icy mind reaches before as well as after the discovery of other forms of intelligent life in the cosmos, right up to the moment when the presumed foundation of its superiority—"naturalness"—is directly imperiled, are the outcome of immaculate analytical operations, deprived of any kind of narcissistic premise which might arise from possible emotional contradictions in its being.

The starting point of the action in "Crusade" is "a certain lack of essential data." The transience and fragility of the world of the giant ammoniac mind—in aeonian proportions, of course—compel it to act to preserve itself. Thus it takes a step that Clarke considers to represent a necessary phase in the development of every cosmic being. A dawning awareness of the entropy that will relentlessly destroy the "icy balance" in which the world of the "natural computer" rests, and precipitate the planet towards the red-hot cores of the galaxies, demands that envoys be sent out into the cosmos in search of "comrades in intelligence," which might have already faced this problem earlier and have found a solution.

However, the envoys establish that similar types of entity are not prevalent in the universe, but find an almost completely opposite form of intelligence, a nonelectronic, "warm" one. This is the key point in the first part of "Crusade." It is precisely this difference, the circumstance that other inhabitants of the cosmos manage to survive in seemingly impossible "warm" environments, that the icy mind fears most, and that provides sufficient reason for trying to make contact with them. This is even more the case because the beings from the "warm" worlds use electromagnetic waves to communicate with each other, and this has enabled the envoys of the icy mind to discover them.

This favorable technical circumstance remains unused, however, and the motives that govern the "natural computer" when it decides not to make contact are especially interesting in the context of our discussion here. The most likely factor in the decision—fear of the inhabitants of the completely different "warm" worlds—has been dismissed in advance, since examination of the recorded data about them has shown unambiguously that they are beings of inconstant structure, short-lived, and with very slow thought processes. These facts enable the icy mind to take upon itself to be guided by the assumption that electronic intelligence is superior to the nonelectronic kind.

Regardless of the reasons that the "natural computer" has in mind when it misses taking the technical opportunity to make contact with nonelectronic intelligence, it does not remain indifferent to it. The natural computer nevertheless establishes attitudes towards the inhabitants of "warm" worlds, but their markedly aggressive character bears unambiguous witness to the fact that these are based on emotional contradictions.

It should not, however, be thought that there exists any inconsistency in the construction of its "psychic portrait." The icy mind still does not display an a priori hostile and intolerant attitude towards alien forms of intelligence; that is, its attitude is not of a xenophobic nature. It insists on directing itself according to the facts, without apparent regard to the strange and unusual nature of those facts. The data it acquires on nonelectronic "warm" intelligence do not provoke this reaction even when it becomes certain that the latter form is considerably more prevalent in the cosmos than "icy" electronic intelligence.

It is only the final data obtained by its envoys which brings down the rampart of indifference around the "natural computer," transforming it into a merciless cosmic inquisitor. Its examination of the signals broadcast by the inhabitants of "warm" worlds points to a fact which immediately threatens to shake the worldview of the icy mind to its foundations. Although assumed to be inferior, nonelectronic intelligence has succeeded in creating electronic intelligence by artificial means and even "in some cases... imposed control" over it.

This "heretical fallacy" brings into question not only the superiority of the icy mind but also its identity. If the assumption that electronic intelligence can be created by artificial means is correct, then, according to the mind's same analytical logic, its status of independent entity is fundamentally disputed, since the condition for "natural" origin is apparently no longer met.

The problem of origin which arises here brings the "natural computer" to complete confusion. Its analytical mind, no matter how mighty, is no longer in a position to break out of its own provincialism and to find a way out of a situation which it almost identifies with the classical scholastic circulus vitiosus of the chicken and the egg.

The only way left to the icy mind to resolve this problem, when all attempts to unravel it "from the inside" fail, is removal of the direct cause of the problem. In defense of its assumed evolutionary primacy or its superiority, the computer embarks on an open "crusade" against those who have had the temerity to bring into doubt the basic principle of its catechesis—its exclusively "natural" origin.

The title of the story has already unambiguously shown the nature of the campaign which the icy mind is undertaking. This title also, however, implies that Clarke has intentionally modeled his central character on the idea of the "cosmic conqueror."

The absence of man from the forefront of the story, and the existence in the story only of an "alien" being which is markedly anthropomorphized, again suggests that the nonhuman protagonist in fact represents a parable of man, as was the case in "Report on Planet Three." This time, Clarke's reason for opting for a change of roles is primarily because by turning man into an alien being in relation to the central character could show the contradictions one falls into when one attempts to preserve, at any cost, one's own presumed superiority, or the illusory and imperiled singularity of "natural" origin.

The fallacy which transforms the objective analytical mind into a blind cosmic inquisitor is based on a conviction in the loss of the status of entity, a status that might possibly have originated in an artificial rather than a natural way. Clarke's fundamental purpose is to show the untenability of the yardsticks for the status of entity which are based on a disproportionate natural/artificial duality.

It is not in the least accidental that he has chosen the nature of the intelligence of the two groups of entities as the key to their difference. Man as the representative of nonelectronic, biological intelligence has, even today, an opportunity to confront directly a completely different type of intelligence, of which the icy mind of "Crusade" is a considerably more advanced form. Our attitude towards this other, electronic, nonbiological intelligence is the same as that of the main character of the story towards the "warm" forms of intelligence. We will remain indulgent towards it right up to the moment when it threatens to bring our superior position into question.

The central character of "Crusade" is not so much worried by the fact that the inhabitants of "warm" worlds have managed to create electronic intelligence artificially, because it has itself managed to reproduce itself, but because its status of entity, based on a conviction in the exclusive "naturalness" of its own being, is thereby apparently disputed. The campaign upon which the "natural computer" embarks represents a particularly anthropomorphic reaction which Clarke purposely clothes in religious attire to make it as obvious and as expressive as possible. This is supported by the dialogue between the icy mind and its envoys in the second part of the story. This dialogue reminds one of a bench of inquisitors making a decision about the fate of "heretics."

It is worth bearing in mind when considering this story that it is in fact about man's attitude towards electronic intelligence, which he has indeed created but which is increasingly slipping out from under his control. Clarke thoroughly brings into doubt the objectivity of man's criteria for the status of entity which are based on the assumption of "naturalness" as a true yardstick.

In this way, an "artificial," electronic intelligence is automatically provided which, Clarke quite rightly considers, does not have to differ qualitatively from "natural," biological intelligence. It is just because of this that the roles have been swapped, because the reader has the chance to perceive the real roots of the fallacy of the icy mind if he knows reliably that the other, nonelectronic form of intelligence could indeed arise by natural means.

The anthropomorphism which Clarke concentrates on in "Crusade" is somewhat more complex in nature and method of action than the anthropocentrism considered previously, and could be designated as anthropochauvinism.

1.2.3 "History Lesson"

Anthropomorphism, as a specific deficiency in the perspective of a human being, appears in yet another form in those of Arthur C. Clarke's science fiction works which deal with the theme of "first contact." In the previous two cases it involved rejection of any possibility of the existence of alien forms of intelligent life, or of allowing that possibility on condition that man's superiority is not imperiled by it. There is this time no doubt not only that alien entities exist but also that they can be superior to humans; however, even this considerable flexibility is still insufficient for their comprehension.

In contrast to the first two types of anthropomorphic deficiency, anthropocentrism and anthropochauvinism, in which the perpetrator in question reveals himself at the level of a priori attitude, the third type, simple anthropomorphism appears as an innate deficiency in man's cognitive apparatus, which is expressed quite independently of any other attitude. A good example of the third type of anthropomorphic fallacy is found in the story "History Lesson."

As in "Report on Planet Three," there are two narrational perspectives, but with the difference that it is now Earthlings who play the part of chronologically older protagonist, although their role within this work is subordinate.

The plot focuses almost exclusively on the chronologically younger protagonists, the Venusians. They are aware of the existence of their Earthling forerunners, whose planet is covered in ice and has long been bereft of any form of life. Immediately before their extinction, however, the last generation of semi-wild descendants of the once highly civilized inhabitants of Earth preserved certain relics for the future, including several items from the post-technological era, items whose meaning they have never attempted to grasp.

Although the Venusians are in this respect more enterprising and persistent, relying on their highly developed science, the outcome is in the end the same. They arrive at the facts scientifically, but their interpretation completely collapses, although the cause is in this case quite different from that in the previous stories.

Contrary to the Martian scientist in "Report on Planet Three," the Venusian historian does not have an a priori hostile attitude towards Earthlings. And, in contrast to the "natural computer" from "Crusade," he not only allows the possibility that the intelligent beings on Earth were radically different from the reptilian inhabitants of Venus, but is also prepared to openly confront the fact that their "remote cousins" had been wiser and superior in relation to the Venusians. Nevertheless, this objectivity and flexibility are insufficient to remove the destructive effect of Venus-centered planetary provincialism which this time appears in its most complex form.

Discovered among the remains of the vanished terrestrial civilization, there is a film which, to the Venusian experts, represents the main clue in their endeavors to reconstruct the culture of an extinct race. An immaculate analytical apparatus is set in motion to ensu re as correct an interpretation as possible of the tiny celluloid pictures which contain the secret of the appearance, psychology, and intellectual achievements of the defunct Earthlings. In order to increase the objectivity of this procedure, the possibility is considered that what is involved is "a work of art, somewhat stylized, rather than an exact reproduction of life as it had actually been on the Third Planet."

All the disagreements start from this point. What the Venusian historian means by "art" is formed by how imaginative expression is conceived of on the second planet of the Solar System. We learn directly from the historian himself the fundamental assumptions of this conception. "For centuries our artists have been depicting scenes from the history of the dead world," he says at the beginning of his lecture, "peopling it with all manner of fantastic beings. Most of these creations have resembled us more or less closely..."

The outcome is unambiguous: the character of Venusian art—and at no time does the otherwise objective historian doubt this—is provincial in essence. There follows an ingenuous and apparently correct analogy, with far-reaching consequences. Assuming, based on Venus's example, that artistic expression always remains emphatically representational, regardless of the degree of alienness of the civilization from which it originates, the Venusian historian concludes that Earth is also no exception in this respect.

What is more—and here the trouble starts—if art is essentially representational even when it is offered the possibility of expressing itself in an area which, by definition, permits the least restrained and most unlimited flight of fancy—and predictions of the morphological particularities of alien races form just such an area—then it is quite in order to suppose that artistic statements which, thematically, remain concentrated on the creator's own race can only have a still more emphatically representational bias.

The Venusian historian therefore concludes that, even if a film found on Earth is a work of art, it is only art insofar as it is partly "stylized," so that it cannot be taken as a completely faithful reproduction of real life. Nevertheless—and here is the final fallacy in this seemingly faultless analysis—regardless of possible minor deviations from purely objective reality, the celluloid document can, in his view, be considered a valid and reliable source of information about Earth. The snag lies in the fact that the work in question is a cartoon film made long ago in the studios of Walt Disney.

No matter how hard they try, the Venusian scientists will never find the right key to interpreting the film, and all the conclusions which they might arrive at will collapse because the initial analogy of the all-valid nature of a work of art as only partly stylized reality is inadequate.

The culturally narcissistic nature of this analogy is apparent precisely in the Venusian historian's inability to break free of the Venusian understanding of art, which he unconsciously generalizes to the level of universal cosmic yardstick. A particular share in this fallacy is taken by the irony that the only extant document which can offer the Venusians basic information about the Earthlings' civilization is a Walt Disney cartoon, that is, a very specific form of artistic expression which in no way corresponds to the manner in which the Venusian scientist sees art.

At first sight, the focus of the story is upon this irony. This is also supported by the story's structure—a movement along a gradually rising line, right up to the climax in the last sentence, when the immediate cause of the Venusian historian's fallacy becomes clear.

But the real causes lie elsewhere. If the focus had been upon the final sentence, the story would be unconvincing. As in the previous cases, it would not provide sufficient motivation for the excessive anthropomorphizing of the Venusian scientists, especially the historian. Only when it is borne in mind that Clarke's basic intention is to highlight the a priori culturally narcissistic nature of all analogies used in the process of drawing comparisons between two unlike entities, during which process they are completely derived from the particularity of one of those entities, does it become clear that, again, there is an intentional exchange of roles involved, and that the whole of the second, focal part of "History Lesson" is directly concerned with man's perspective.

Without the role exchange, the ironic twist at the end would have been impossible; although this does not occupy a focal point in the story, it does nevertheless have an important role. Again moving his lens from general cultural narcissism onto a special kind of anthropomorphism, Clarke intends in the first place to bring into radical doubt man's cognitive apparatus, which relies to a large extent on an analogy that always remains conditioned by anthropomorphic viewpoints.

As the parable of an Earthling scientist, the Venusian historian does not make a conscious mistake when he places an equals sign between the conception of art attributed to the two planets. This analogy is a reflection of the special character of his way of thinking which, in spite of its undoubted flexibility and objectivity, nevertheless remains, in the last analysis, distorted by "human" yardsticks.

What directly emerges from this conclusion is not exactly rosy for man. He is, namely, capable of the simple gathering of facts (on the level of phenomenon), and from this point of view the requirement of "scientific objectivity" is mainly satisfied. However, when he moves on to synthesizing and interpreting these facts (the level of noumenon), anthropomorphism comes without fail into play as a powerful limiting factor. This is manifested in the range from an exclusively anthropocentric attitude, through the somewhat milder representation of anthropochauvinistic superiority, to the characteristic anthropomorphic restrictions of man's cognitive apparatus.

1.3 "A Meeting with Medusa"

1.3.1 "There Is Life on Jupiter: And It's Big..."

Our consideration of examples of Clarke's stories which use the alien as a mirror in which to see ourselves clearly shows that he is well acquainted with the essence of the problems of anthropomorphism. It will therefore be especially interesting to examine how he tries to supersede its disintegrative action from that of the narrative voice or rather stories where the alien entities have an independent status which precludes their anthropomorphization.

For analysis in this direction, we have chosen the novella "A Meeting with Medusa" because, from the point of view of the conception and presentation of an alien protagonist, and of the examination of the ultimate frontiers of prose narration within a special type of first contact, it is the most famous work of Clarke's sf opus, and undoubtedly ranks among the most successful in the science fiction tradition generally.

In the novella "A Meeting with Medusa," there are two alien protagonists in addition to the human ones, but (in complete accordance with Clarke's basic intention) it remains uncertain up to the end whether they are indirect entities or some transitional form between this status and that of nonentity. The work is divided into two major scenes, one on Earth and a second, which is considerably longer, located on Jupiter several years later. The two parts are linked by the same central character, Howard Falcon. At the end of the first part, after surviving a catastrophic accident, he becomes a cyborg, a special symbiosis of man and machine, that is, a being who is no longer exclusively anthropomorphic (that is, not totally human)—but the reader learns of this change only at the end of the novella.

The fact that the episode that takes place on Jupiter is given much more space in the novella is a reliable indicator that the author is giving it much greater weight than the part that takes place on Earth. Superficially, "A Meeting with Medusa" is about Man's first mission to the largest planet of the Solar System, a mission with the main task of solving certain exophysical puzzles of that gigantic world. The probes which have been dropped earlier into Jupiter's atmosphere are no longer suitable, because it is now not simply a question of merely gathering physical and chemical data but of a more complex form of investigation which requires direct human presence.

However, Man will not show himself to be completely equal to this mission, not so much when it comes to understanding the exophysical characteristics of a world so very different from Earth but when the possibility arises that that world could contain some forms of life. Although no one has seriously expected such an encounter before the mission, there did exist a certain preparedness for that possibility. This precaution is all the more significant for our study because it presupposes certain criteria which can help to determine what constitutes a living being.

The first hint of these criteria is given by one of the characters of the second part of the story, the exobiologist Dr. Brenner: he thinks that the phenomenon of life represents not the exception but the rule in the universe. However, his cosmic diffusion of life is limited by various natural environments to the level of proportionately simple organisms, while one can only guess at the more complex ones.

Considering the possibility of the existence of certain forms of life on Jupiter, Dr. Brenner concludes: "I'll be very disappointed... if there are no microorganisms or plants there. But nothing like animals, because there's no free oxygen. All biochemical reactions on Jupiter must be low-energy ones—there's just no way an active creature could generate enough power to function."

The exobiologist has taken as his yardstick of life the evolutionary model found on our own planet. This model could possibly be valid elsewhere at lower levels of development, and Dr. Brenner is right not to exclude the possibility that certain microorganisms might be found on Jupiter and even some simple equivalent to plants.

His conclusion that the absence of free oxygen on Jupiter means that beings which might correspond in terms of level of development to terrestrial animals cannot exist there, rests on a mistaken belief that the chemistry of oxygen, on which earthly life is based, is universally valid. We met this same type of fallacy, in its intentional form, in "Report on Planet Three."

This fallacy will soon be unmasked. Falcon discovers certain forms of life that he assumes to be considerably more complex than microorganisms and plants. However, a further implication arises from Dr. Brenner's statement. Not for one moment does the exobiologist bring into question the possibility of there being a difference between living beings and the non-living phenomena in the atmosphere of Jupiter. This differentiation is based exclusively on size. His point of view is soon confirmed by Falcon in the Kon-Tiki space capsule: after looking through his telescope, he declares that

"[There] is life on Jupiter. And it's big..."

"The things moving up and down those waxen slopes were still too far away for Falcon to make out many details, and they must have been very large to be visible at all at such a distance. Almost black, and shaped like arrowheads, they maneuvered by slow undulations of their entire bodies... Occasionally, one of them would dive headlong into the mountain of foam and disappear completely from sight."

The standards by which Falcon judges his discovery of living beings in Jupiter's atmosphere are obvious ones. They involve a demonstrable aspiration towards purposeful, meaningful "behavior," manifested in this case as a regular rhythmic movement which cannot simply be the product of the blind and chaotic forces of nature, but must be the result of a certain organization of a higher order. Although the reasons for this "behavior" do not have to be intuitively evident, it always has its phenomenal, discernable aspect, through which unarticulated natural phenomena can be perceived in the background.

However, there may appear in nature nonliving phenomena characterized by hints of similar meaningful and purposeful "behavior." A good example of these phenomena in "A Meeting with Medusa," the gigantic "Poseidon's wheels," are an exceptionally law-abiding light phenomenon which at first makes Falcon think that there are living beings in front of him.

With similar nonliving natural phenomena, however, the noumenal background can always be easily comprehended: Mission Control very quickly discovers the key to this unusually regular fiery display in the Jovian atmosphere on the basis of corresponding phenomena from the oceans of Earth. In the field of the non-living, there are no noumenal differences between the phenomena: the "Poseidon's wheels" will in principle be the same both on Earth and on Jupiter.

The differentiation starts only on the level of life, because here conclusions can no longer directly be drawn intuitively on the basis of phenomenon. This split between the level of phenomenon and that of noumenon is not, indeed, significant at lower degrees of evolution, where the noumenal identicalness of natural phenomena is still proportionally preserved. Things, however, change radically with the appearance of organisms which possess self-awareness.

The central character of the novella is first confronted with the difficulties which appear in relations between the planes of phenomenon and noumenon when he tries to understand something more of the nature of the living beings he encounters first in the Jovian atmosphere. Falcon establishes that they are creatures far larger than any earthly ones, which is not strange when one bears in mind that they are made according to the measure of the world they inhabit. Closer examination shows him that these unusual creatures have nothing that might remind him of sense organs—and this is also understandable, considering that every similarity with terrestrial creatures on the plane body structure would be in obvious disharmony with the great exophysical differences between the two planets.

In both these cases, Falcon does not succumb to possible fallacies of anthropomorphism. Without reluctance, he readily accepts the possibility that the proportionately evolved beings living within the gaseous mantle of the giant planet are essentially different from the inhabitants of our own world, both in shape and in size.

Problems arise when he needs to fathom those specific characteristics of the mantas that cannot be identified through simple observation. Falcon tries to discover some higher order in the "behavior" of these creatures which might help him to discover the possible purposefulness directing them, the key to their "intelligence." But he suddenly comes up against a dead end because the available data by which he might arrive at some reliable pointer to the noumenal nature of the huge creatures living in the clouds of Jupiter are shown to be either insufficient or ambiguous.

It turns out that the secretive mantas can be either unintelligent, harmless herbivores or intelligent bandits. Since they pay no attention to the Kon-Tiki during their first encounter, Falcon at first concludes that they are indeed harmless vegetarians. The events of the next day cause Falcon to change his opinion: these same mantas, which had completely ignored him while he floated among them, simply change into intelligent bandits with a highly developed strategy of attack when they pounce on those other strange inhabitants of the Jovian atmosphere—the giant medusae.

It is symptomatic that, in both cases, there is the same measure of intelligence: a capability for aggression. Falcon's initial conclusion that the mantas are not intelligent is based on the fact that they do not attack him, while their transformation into "intelligent birds of prey" is directly conditioned by the circumstance that they take an aggressive attitude towards the medusae.

Doubts nevertheless remain with regard to the possibility of establishing the intelligence of the mantas only on the basis of their external "behavior," in view of the fact—as is soon demonstrated—that the attack on the medusae was fated to fail from the beginning, because the victim, not so fated, has a weapon which would discourage a far mightier and more intelligent enemy, and it is evident that there are certain contradictions in the "criterion of aggression" which Falcon had in mind when coming to the above conclusions.

The nature of these contradictions becomes clear if one considers more closely the name which Falcon gives to these strange "mantas." At first glance, it might seem that he was led to choose this appellation because of the similarity of the form and way of movement of these strange inhabitants of the great waxen clouds to that of manta rays. The events in the first part of "A Meeting with Medusa" suggest, however, that this seemingly superficial analogy has considerably deeper roots.

1.3.2 Medusae and Mantas

The tragic crash of the giant dirigible, the Queen Elizabeth, indirectly enables Falcon to become, as a cyborg, a suitable astronaut for the mission to Jupiter. At the same time, his human identity is seriously brought into question. During the Jupiter episode, the disunion between the "nightmares brought from Earth" and the new, no longer human status to which he increasingly belongs reaches a culmination.

The Queen Elizabeth resembles an inhabitant of the seas of the planet Earth which in its resembles a jellyfish: a medusa.

"He had once encountered a squadron of large but harmless jellyfish pulsing their mindless way above a shallow tropical reef, and the plastic bubbles that gave Queen Elizabeth her lift often reminded him of these—especially when changing pressures made them crinkle and scatter new patterns of reflected light."

The association is, at this moment, a completely spontaneous one, and there are no complex themes behind it at all. However, each time it reappears, even if only in an indirect form, it is burdened with references to the tragic events that follow soon after its first appearance. In Falcon's nightmares, indeed, the past happenings are not so much linked with the air crash itself as with the moments and hours after regaining consciousness—his rebirth. But the last, firmly rooted representation from his previous, human status of the associations of the Queen Elizabeth will, like a gigantic medusa, acquire the value of a double-meaning symbol, the ominous nature of which will change depending on which of Falcon's two identities—innate and human or acquired and cyborg—predominates.

Although the conflict between these two identities started during his physical recovery on Earth, the exceptional circumstances in which Falcon finds himself while descending through the atmosphere of Jupiter are intensified to the utmost limit. This intensification has, however, a gradual character: the initial circumstances much more stimulate fear of the loss of his old identity than joy in acquiring a new one.

It is quite understandable why the encounter with a possibly intelligent entity in medusa form cannot arouse euphoria in him. It awakes recollections of a completely different kind.

When he calls the strange inhabitants of the gigantic waxen clouds "mantas," Falcon defines his attitude towards them, casting doubt on the validity of his conclusions about the nature of these creatures and practically preventing him from developing any intuition about them. Conditioned by feelings of danger and fear, Falcon's perception of the mantas narrows down to the plane of aggression, and this inevitably results in the anthropomorphization of aliens by ascribing to them a negative emotional stance towards man.

Only with this in mind can we understand the background to some of Falcon's statements during his encounter with the mantas. For example, the effect of his attitude is evident in Falcon's first statement after he has informed Mission Control of his discovery of living beings. Up to this moment, Falcon has been at a safe distance from the mantas, and that they have been paying no attention to him. "And even if they try to chase me," he says, stifling the echo of a distant earthly cry, "I'm sure they can't reach my altitude."

The next day, while he is watching a shoal of mantas charging an enormous medusa, Falcon abruptly declares this move to be an attack, but soon realizes that the facts do not favor such a conclusion. Above all, the differences in the sizes of these creatures are so great that the mantas on the back of the medusa appear "about as large as birds landing on a whale." When the medusa reacts to their presence, Falcon immediately returns to his first instinctive assumption and even identifies emotionally with the "attacked" medusa.

"It was impossible not to feel a sense of pity for the beleaguered monster... Yet he knew his sympathies were on the wrong side. High intelligence could develop only among predators—not among the drifting browsers of either sea or air. The mantas were far closer to him than was this monstrous bag of gas."

The easy and effective defense by the medusa shows that Falcon's intuitions about the mantas rests upon anthropomorphization—an anthropomorphization rooted in fear of the ray—like form of the mantas, and ultimately, in fear of loss of human identity. Aggression as a "yardstick of intelligence" does not help Falcon to perceive the true nature of the bizarre denizens of the waxen clouds.

The medusa's reaction follows too late to remove this yardstick completely. In the meantime, it has even expanded its reach into that area where no direct association with a medusa exists. A link between fear and intelligence also appears between the two encounters with the mantas, when the stupendous firework display of "Poseidon's wheels" begins in front of the astonished Falcon.

Faced with the enormity and regularity of this fantastic natural phenomenon, he conceives for the first time that there might be intelligent beings in the atmosphere of Jupiter. "No man could look upon such a sight without feeling like a helpless pygmy in the presence of forces beyond his comprehension. Was it possible that, after all, Jupiter carried not only life but also intelligence? And, perhaps, an intelligence that only now was beginning to react to his alien presence?"

The possibility of the appearance of intelligent aliens at the beginning of the mission to Jupiter is accompanied every time by a deep feeling of fear. The perplexity that remains after the disappearance of the mantas is, however, properly recompensed by the appearance of a new creature which—at least at a superficial narrative level—shows not only convincing signs of intelligence but less indifference. This encounter with the medusa takes place under circumstances which have an important influence on all the later conclusions that Falcon reaches about this strange creature. This event follows immediately upon the discovery of the mantas—that is, after a specific anthropomorphic mechanism has already been activated in Falcon's consciousness.

Although brief, the events that happen from the moment of the sighting of the huge "oval mass," at the base of a terraced layer of Jovian clouds, until the dusk prevents further observation, are sufficient to determine the direction Falcon's later deliberations on the medusae.

The "oval mass" reminds Falcon of a "forest of pallid trees," since he discerns something resembling "hundreds of thin trunks, springing from the white waxy froth." The lyrical charge that characterizes this association testifies that it is not a question of a simple analogy of notions deprived of any emotional stance, but rather a complex mechanism behind which there no longer stands an indifferent objectivity.

This subdued, emotionally colored image, without precedent in Falcon's earlier mental reservations, suggests that his ability to come to unbiased conclusions is impaired. The nature of this impairment is soon defined by the second image that comes to Falcon's mind: the "oval mass" reminds him of a "giant mushroom"—in which one can already perceive an approximation to the central symbol of his nightmares, that of the "medusa."

The disturbance of the equilibrium of indifferent objectivity here is, indeed, still an inconspicuous and innocent one, since the first passing glance at the "oval mass" has not provided any basis for assuming that a certain form of life is involved; however, when specific indicators suggest this possibility, that equilibrium will be brought into question more seriously.

As in the case of the mantas, the indicator of life in support of the unarticulated laws of nature is represented also this time by a certain coherent organizational order which is not met in non-living phenomena in the macro world. Just before he dives into the shadow of the Jovian night, Falcon sees the incredible synchronization of the strange "trees" bending, which casts doubt on his previous assumption about the nonliving nature of the "oval mass."

In favor of the new assumption that this is a living being is the circumstance that the "enormous tree" is no longer in the same place where Falcon saw it first. Two data are thus learned on the plane of phenomenon that are conditionally relevant for drawing a conclusion as to whether it is a living creature, but are utterly insufficient to learn anything at all about it on the level of noumenon.

Nevertheless, Falcon joins unawares in one such understanding, and the far-reaching, distorted effect of this will seriously affect the validity of his next conclusions about the medusa. Along with the observation that the "oval mass" is a living being, Falcon again links an image that brings into even finer focus the source of his associative course, which concurs with the direction of the previous one. The sight of the immaculately synchronous rhythmical waving of the huge "forest" reminds him of "fronds of kelp rocking in the surge."

The meaning of this idea is obvious: it merges doubly with the image of the "giant mushroom" from the previous association, which is very similar in form to a medusa. On the one hand, the new association defines the location of the central symbol of Falcon's fear—the sea—while on the other, the image of the bending of strands of seaweed directly suggests the sight of "jellyfish pulsing their mindless way above a shallow tropical reef" which has one of the key places in the first part of the novella.

It is quite certain that these two scenes are not only joined by formal similarities but also by a complex referential link that will become evident during the next encounter. When he sees the "oval mass" again the next day, Falcon needs only a few moments for all his previous doubts as to its identity to disperse. The image that now flashes through his consciousness is congruent with the one that occurred to him many years ago on Earth while he was watching the inflating and deflating of the bubbles on the dirigible Queen Elizabeth. "It did not resemble a tree at all, but a jellyfish—a medusa, such as might be met trailing its tentacles as it drifted along the warm eddies of the Gulf Stream."

Here at last we see explicitly how, step by step, Falcon's image of potential intelligent beings gradually expands around the medusas—from the bizarre mantas, by way of the puzzling oval mass, right up to the direct incarnation of the medusae themselves. Falcon immediately takes a negative emotional attitude towards these possible entities, conditioned by fear of loss of human identity that, as we saw in the example of the mantas, casts seriously doubt on the possibility of getting to know them on the level of noumenon.

The manifestation of this negative emotional determinant, that is, the fear that is closely linked with the process of "medusation," also occurs this time, at two characteristic places, immediately after Falcon has reliably established that the medusa certainly represents a higher form of life. In the first case, the fear is evinced in his instinctive use of atmospheric circumstances to justify avoiding approaching the medusa so as to observe it in as much detail as possible. Indeed, the adjective "secure" that he uses to describe his position at that moment could in principle have two meanings: secure from sinking into the lower layers of the atmosphere, and secure from possible arrival within reach of the medusa.

(To go down would present easily predictable exophysical dangers. A certain hesitation, however, in his use of "secure" tells us that what is involved is avoidance of something that arouses in Falcon's consciousness much greater suspicion of the relatively easily predicted exophysical dangers with which he would have been confronted had he gone down to the foot of the terraced clouds.)

The real nature of this suspicion soon surfaces, although there has again been no very serious motive for it on the level of phenomenon. Observing the medusa for some time through a telescope, Falcon suddenly begins to ask himself whether its inconspicuous color is not some kind of camouflage: "Perhaps, like many animals of Earth, it was trying to lose itself against its background. That was a trick used by both hunters and hunted. In which category was the medusa?"

The question is, obviously, just a formal one, because the whole of the previous structure has been erected to suggest only one answer. This answer has already been present, in advance, in Falcon's consciousness, and it was only necessary to provide a convenient occasion for it to be made concrete through some external characteristic of the medusa.

Fear has not for a moment been absent from Falcon's consciousness; it would appear without fail whenever data at the level of phenomenon even conditionally allow it. Everything up to the battle between mantas and the medusa has not really provided a serious motive for this manifestation; a sharper expression of fear before this would have seriously conflicted with the known data on the medusae, that have in this sense been, almost without exception, strictly neutral.

Falcon's consciousness has only been latently, and not maniacally, burdened with one of the strongest phobias that, with inessential differences of degree, is present in all people. If Falcon had been conceived as a psychologically disturbed person who projects the key symbol of his mania everywhere (the medusa-like form), the degree of misconception about the known data on the gigantic inhabitant of Jupiter's terraced clouds would have been much greater.

However much fear prevails in Falcon's consciousness, though, it never gains a pathological dimension; that is, it never reaches a point of confrontation with the knowledge acquired on the level of phenomenon, but arises only when he attempts to pass to the level of noumenon.

That it is not a question of individual disturbance but of a characteristic of human consciousness in general—which is in Falcon's case over—accentuated by the fact that he finds himself in a completely unknown environment. For the first time he encounters heterogeneous forms of life, and in the physical sense, he long ago ceased to be a real man. The complex psychological changes this produces are best shown by the sudden and seemingly unexplained attitude of the astronaut towards the medusa at the moment when it looks as though the mantas' attack has got it into serious trouble.

Only a few moments before, Falcon was taking an explicitly negative emotional attitude towards the creature, conditioned by a subconscious fear of the loss of human identity, but he now suddenly starts to feel sympathy and share in its trouble:

"It was impossible not to feel a sense of pity for the beleaguered monster, and to Falcon the sight brought bitter memories. In a grotesque way, the fall of the medusa was almost a parody of the dying Queen's last moments."

The roles remain unchanged, and only for a moment does Falcon's emotional attitude change—and that only towards the medusa. The strategically well—conceived attack by the mantas which, for a short time, suggests the existence of intelligence, causes a defense mechanism of fear to be strongly activated in the astronaut in relation to these bizarre creatures, a fear which—as we have seen—has occasionally been a little subdued but not completely removed as far as they are concerned. It is therefore only a matter of a change in intensity within the context of the same emotional attitude, not of a change in that attitude itself.

Such a change is arrived at only in relation to the medusa which begins to arouse, instead of a feeling of fear, a quite short-lived feeling of sympathy and inclination. This ambiguity in Falcon's emotional attitude towards the creature can only be explained from the angle of a different, considerably broader ambiguity in the complex being of Falcon: this relates to the above-mentioned split between the strongest fear—the fear of losing human identity—which dictates a negative emotional attitude towards the medusa, and Falcon's endeavors to get used to his new, nonanthropomorphic identity of cyborg, in which context the emotional weight of the medusa symbol is diametrically changed. It now becomes a synonym for a new birth, and the only possible emotional attitude towards it is a positive one.

This emotional change towards the medusa does not arise in an ad hoc and unmotivated way but originates in an event which took place directly before the battle between the inhabitants of the terraced clouds of Jupiter. A seemingly innocent remark by Dr. Brenner has for the first time brought into focus—although not yet fully explicitly—the slow but steady transformation of Falcon, who is gradually and by no means painlessly alienating himself from his human origins in order to get used to his new status of cyborg.

Understanding the reasons that have led Falcon to avoid approaching closer to the medusa, exobiologist Brenner uses the pronoun "we" to express his consent to Falcon's wish to retain his present altitude. However, "that 'we' gave Falcon a certain wry amusement; an extra sixty thousand miles made a considerable difference in one's point of view."

The difference which has here been expressed in units of spatial distance will change at the end of the novella into the fundamental and unbridgeable difference between human beings on the one hand and the cyborg Falcon on the other. Nevertheless, however much this first hint of that all-embracing transformation has been superficial and subdued, only it can be a valid motivation for the short-lived ambiguity of Falcon's emotional attitude towards the medusa. Minutes later, this ambiguity is soon resolved when the medusa uses its "secret weapon," in favor of a strong tide of fear. But in the distant future, this ambiguity will make Falcon more capable of the act of contact with alien creatures. During his first mission to Jupiter, he is not mature enough to make contact, still too overwhelmed by the sharp contradiction of his imperiled anthropomorphic status.

1.3.3 Prime Directive

The spectacular counter-attack by the medusa, which disperses the mantas forever from the scene of events, reveals a further significant feature of this strange creature. It is revealed that the "monstrous bag of gas" has an organ that acts like a special kind of radio aerial. The possibility that the medusa possesses a radio-sense arouses two particular types of reaction: phenomenal and noumenal.

Dr. Brenner's opinion on this unusual organ does not arise from the immediate external functionality of radio aerials in the special biophysical environment of the Jovian atmosphere. Starting from the assumption that senses arise depending on the prevailing physicochemical stimuli of a given world, the exobiologist concludes that it is not at all strange that the radio-organ has not developed in any terrestrial organism, as it would have been superfluous in the biophysical conditions of our own planet. There is, however, radio energy from Jupiter in abundance, and there it represents a very important factor in the physical environment, so that evolutionary processes could not simply neglect it.

Looked at from this angle, the medusa's radio sense is no less probable than the human eye, terrestrial evolution's response to the amount of light radiation which prevails on Earth. The conclusions that Dr. Benner draws from this fact do not overstep the limit of the level of phenomenon (just like the conceptual characteristic which Clarke, in his role of "omniscient narrator," ascribes to the medusa): "Until I came here... I would have sworn that anything that could make a short-wave antenna system must be intelligent. Now I'm not sure."

In other words, a phenomenal characteristic which would, in the context of the human world, undoubtedly point to an artificial origin and to the existence of intelligence, does not have prove it in a radically different environment where the conditions exist for it to develop in a natural way. The exobiologist's reluctance to interpret the presence of the radio aerials as a reliable sign that the medusa is intelligent (which represents a kind of progress in relation to the previously mentioned and principally anthropocentric standpoint of Dr. Brenner), corresponds to an avoidance, on the basis of data from the plain of phenomenon, of bringing a judgment by direct analogy on the level of noumenon, which would in this case have anthropochauvinistic characteristics.

But, however much this standpoint seems to be right at first sight, it nevertheless has one serious deficiency: the scope of its validity is rather limited. Keeping to the level of phenomenon when encountering alien beings can be sufficient only up to the moment when the situation produces a possibility of making contact.

Contact presupposes the existence of sentience, a completely noumenal characteristic that—as we seen in the examples of Dr. Brenner's conclusions—cannot be learned by simple observation of alien beings on the level of phenomenon. It is clear that the transition to the level of noumenon is necessary. Of course, this necessity by no means guarantees that contact is possible.

All Falcon's attempts in that direction have, up to now, been a failure. In contrast to Dr. Brenner, who does not move from the level of phenomenon, Falcon tries on several occasions to draw conclusions on the plane of noumenon, but each time he is thwarted from doing this by anthropomorphism. Indeed, in circumstances which might conceivably open up the possibility of making contact, there is no way in which it can be reliably concluded that the medusa is aware of the presence of the astronaut, and noumenal conclusions cannot have any important influence on Falcon's direct actions. Thus, for example, his supposition that the medusa is using its radio-sense to monitor communications between Mission Control and the Kon-Tiki, that is, the hint that it is a highly intelligent creature (a conclusion diametrically opposed to that of Dr. Brenner) remains without any echo on the plane of direct action.

A certain change does nevertheless take place, in view of the fact that even the possibility of an encounter with an intelligent creature is sufficient for the Mission Commander to order Falcon to be guided as a precaution by the "Prime Directive." These special instructions for first contact have originated mainly on the basis of man's experience on Earth. Unfortunately, Falcon comes to realize that the rules have a basic deficiency stemming directly from the one-sided nature of human experience on our planet.

The Prime Directive is founded on the assumption that humans will be the only type of participant in the act of making contact. "For the first time in the history of space flight, the rules that had been established through more than a century of argument might have to be applied. Man had—it was hoped—profited from his mistakes on Earth. Not only moral considerations, but also his own self-interest demanded that he should not repeat them among the planets. It could be disastrous to treat a superior intelligence as the American settlers had treated the Indians, or as almost everyone had treated the Africans..."

The compilers of the Prime Directive have failed to emerge from the framework of terrestrial experience, where Man has the opportunity to meet exclusively with homogeneous races which differ only in terms of insignificant morphological characteristics and degrees of civilizational development, and have simply remained blind to the possibility of encountering essentially heterogeneous entities which need not have any common noumenal denominator with the human race. All the other defects in the rulebook's provisions have originated from this classic contradiction.

The incomplete and limited nature of these provisions is especially evident when the method and scale of direct action at the moment of making certain forms of contact has to be determined. The first clause of the Prime Directive already contains a by no means innocent vagueness which gains special weight when Falcon is making his decision on how to react to certain actions by the medusa that could, but do not have to be, interpreted as an initiative for making contact.

This first provision states: "Keep your distance. Make no attempt to approach, or even to communicate, until 'they' have had plenty of time to study you." While the first part of this instruction matches the attitude which Falcon—from totally different motives—takes towards making contact with the medusa, the second part arouses serious doubts.

"Exactly what was meant by 'plenty of time,' no one had ever been able to decide. It was left to the discretion of the man on the spot."

Unfortunately, as far as the medusa is concerned, Falcon's decisions in this area have long ago lost the characteristic of unbiased objectivity, and he is swayed by anthropomorphic factors.

Nevertheless, before Falcon finds himself directly testing the validity of the provisions of the Prime Directive, the anthropomorphic mechanism needs to be subdued for a second time. At the start of the last act of the drama on Jupiter, Falcon suddenly gains the paradoxical insight that he, who is physically no longer a human being, might well become the "first ambassador of the human race."

This feeling appears to reflect a hidden, broader contradiction in Falcon's double identity and seems likely to produce change of attitude towards the medusa. However, Clarke prevents this by concentrating on the lethal weapon that the "monstrous bag of gas" has available. The occasion for pointing his thoughts in this direction is provided by the peculiar atmospheric conditions on Jupiter that bring the space capsule nearer and nearer to one of the medusae. Although he presumes that the range of its defense mechanism is rather limited, Falcon does not at all wish to get involved in personal investigation.

However, more important at this moment than the renewed current of fear, the roots of which are quite clear, is the fact that, for the first time, Falcon thinks of the possibility of direct action in relation to the medusa. "The wind that was steadily sweeping Kon-Tiki around the funnel of the great whirlpool had now brought him within twelve miles of the creature. If he got much closer than six, he would take evasive action." This hint that Falcon would, in principle, avoid getting any closer to the medusa—his readiness to act to nip any such possibility in the bud—has a special place in the events which soon follow, events to which Falcon brings a whole series of contexts for our consideration.

1.3.4 Noumen and Phenoumen

The events described in the chapter "Prime Directive" are different from those we have considered up to now because they suggest that, for the first time, both protagonists in the encounter are aware of each others' existence. Right up to the moment when one of the medusae unexpectedly appears immediately above the space capsule, only the human was reliably aware of the encounter with the alien. The mantas and more distant medusae showed no signs at all of having noticed the tiny earthling spacecraft. Unfortunately, there are very meager data on the circumstances which nevertheless caused one medusa to "spot" the alien in its world.

For this occasion, Clarke suddenly introduces a constructional defect in the Kon-Tiki: its large silver balloon prevents the area above the craft being inspected either optically or by radar. This unfavorable technical circumstance allows Falcon and the strange inhabitant of the terraced clouds to find themselves in direct proximity, enabling the medusa to "react" to the Kon-Tiki's presence.

In fact, except for mere chance, Clarke has no other option at his disposal: this momentous convergence could not have happened on Falcon's own initiative without contradicting the essential features of his psychological makeup presented so far.

On the other hand, if the initiative had been ceded to the medusa, Clarke would have made a big mistake with regard to the logical coherence of his novella: he would, in his role of the omniscient storyteller who conceives his heroes and predicts their actions, have stepped much further onto the plane of the noumenon of the strange inhabitants of the Jovian atmosphere. To do so would seriously impair the basic assumption underlying the type of storytelling demanded by the theme of "first contact."

All the subsequent events follow as the outcome of certain "actions" by the medusa, but there will be no foundation for concluding more reliably or in greater detail their real nature. Regardless of this uncertainty, however, there are two significant circumstances which force Falcon to react in a specific way to the particular "initiative" of the "monstrous bag of gas," even though the provisions of the Prime Directive are sharply opposed to such reactions.

Above all, at the moment of direct encounter, Falcon's fear of losing his human identity is again aroused by the particular circumstances that precede the appearance of the medusa above his capsule: Before he discovers the actual presence of the medusa in his vicinity, he is faced with a dramatic, though indirect, hint of its appearance, for which he is at first unable to find any explanation. The gigantic "oval mass" that looms above his craft causes darkness to fall suddenly over the surrounding area, even though there remain several more hours to sunset.

This unpleasant optical puzzle soon receives acoustic clues. From some-where out of the immediate vicinity—and no longer through the radio link—Falcon begins to hear without any warning or announcement a spectral, cacophonous crescendo of medusa "noise" with which "the whole capsule vibrated like a pea in a kettledrum." This rise in tension finally removes the last ambiguity from Falcon's attitude towards the "monstrous bag of gas," leaving room only for overwhelming fear.

The second important circumstance influencing the direction of Falcon's actions and linked to his emotional state is the very vagueness of the medusa's actions, which allow contradictory interpretations.

The process which we can conditionally designate as an "attempt to make contact" between the medusa, as the "initiator," and Falcon, who is reacting to this "initiative," is in four separate phases, the common denominator of which is the identical nature of the actions of the "monstrous bag of gas": that is, it makes a special kind of "approach" in which Falcon primarily perceives a reliable indication that the medusa is aware of his existence. The first phase begins when a fence of thick tentacles suddenly descends around the capsule, the climax of the tension that has been growing ever since the surprise eclipse.

In the process of responding to this initiative by the medusa, Falcon finds himself choosing between two principal models of reaction: the noumenal and the phenomenal.

The first model specifies his response on the basis of a particular interpre-tation of the action of an alien creature—an interpretation that implies the possibility of noumenal knowledge of the medusa. The second model does not attempt to comprehend the alien's intention, and thus does not specify the direction of response but confines itself to establishing what should not be done. This is not, however, because that sense could not possibly be reached a priori, but because it contains too many unknowns to establish the actions to be taken. Here, the problem of getting to know an alien being on the level of noumenon is not posed at all. Falcon's consciousness, dominated by fear, corresponds to the noumenal model, while the phenomenal model is embodied in the Prime Directive.

Although both these models are available to Falcon immediately before the medusa's action, he does not make a choice based on a process of sensible elimination but instinctively opts for the possibility that has prevailed in his consciousness up to now and that envisages a specific response. In deciding with a "lightning-swift movement" to pull the rip-cord of the balloon and thus, without aforethought, to escape the embrace of the medusa's tentacles, Falcon simply succumbs to the effect of renewed and strengthened fear. From this perspective, the medusa's action could only be interpreted as aggressive and could only have been responded to by a hasty retreat.

The later events on Jupiter also remain completely part of the noumenal model of Falcon's reaction to the initiative of the "monstrous bag of gas," and this results in their outcome being the same as in the first case. Nevertheless, although he always responds to any approach by the medusa by retreating, he is twice in a position to react from the point of view of the phenomenal model.

The first such occasion arises in the second phase of the "attempt to make contact." The medusa "responds" to Falcon's escape by coming still closer, but then suddenly stops at a distance of less than a mile above him. It is only this complete cessation of activity by the creature that allows Falcon to come out of his cocoon of fear and consider the medusa's initiative from another angle.

Falcon's reflection on the causes which might have prompted the bizarre inhabitant of the Jovian atmosphere to keep itself at a certain distance includes both the noumenal and phenomenal models which were previously available when he attempted to learn about the strange creatures. "Perhaps it had decided to approach this strange intruder with caution; or perhaps it, too, found this deeper layer uncomfortably hot."

In the first interpretation, the medusa's activity is interpreted from a noumenal perspective, as shown unambiguously by the words "decided" and "with caution." The second, phenomenal interpretation does not pretend to penetrate to the possible inner motives for the medusa's action, but is limited to external physical explanations. Although this phenomenal interpretation is seemingly more reliable than the noumenal one, it is significant that it appears only when the medusa is not attempting to make contact.

Confronted with such a limited field of action, Falcon remembers Dr. Brenner's warning about the provisions of the Prime Directive. At that moment, Falcon recalls a television discussion between a space lawyer and another astronaut:

"After the full implications of the Prime Directive had been spelled out, the incredulous spacer had exclaimed: 'Then if there was no alternative, I must sit still and let myself be eaten?' The lawyer had not even cracked a smile when he answered: 'That's an excellent summing up.'"

Here, the lack of a phenomenal model of behavior towards aliens during first contact is finally made concrete. Because the Prime Directive limits itself to giving instructions exclusively about the things which should not be done—and first contact intrinsically requires some action, if only in responding to the initiative of the other party—it is obvious prescription on the level of phenomenon is insufficient.

Its value remains until the need for direct action arises, after which it is unable to offer any kind of plan. In this it differs from the noumenal model that does indeed suggest a certain plan, but its validity remains completely overshadowed by the distorting effect of anthropomorphism. The final abandonment of any attempt to reciprocate using the phenomenal model follows immediately in the third phase when the medusa again goes into action. This time its initiative is most clearly an "attempt to make contact," but Falcon's fear has already caused him to ascribe an a priori aggressive nature to all "overtures" by the opposite side.

This is best shown by the association aroused in Falcon's consciousness by the unexpected elongation and descent of one of the medusa's tentacles towards the Kon-Tiki. "As a boy he had once seen the funnel of a tornado descending from a storm cloud over the Kansas plains. The thing coming toward him now evoked vivid memories of that black, twisting snake in the sky."

Whatever it actually represents, this step by the medusa means only one thing to the astronaut: an attack, directly manifested through an assumed attempt by the creature to "eat up" the Kon-Tiki. Falcon response is thus to "frighten" the bizarre Jovian inhabitant so he can gain time to retreat. But since this retreat has to be a final one, he hesitates for a moment, waiting for Dr. Brenner's opinion. And now follows the key moment that at last gives practical confirmation of the inappropriateness of the Prime Directive to the purpose for which it was intended. Both before and during the medusa's initiative, the exobiologist has remained strictly faithful to the phenomenal model of reciprocity. Now, when he finds himself for the first time in a situation where he cannot simply decide what should not be done but must propose some direct plan of action, he has nothing to say to Falcon's decision to withdraw even though it is contradicts the Prime Directive.

After the decision to withdraw becomes irrevocable, it seems to be the end of the drama on Jupiter. The specter of fear has prevented Falcon from responding to the medusa's initiative in any other way than by retreating. And indeed, he immediately starts up the engine ignition sequence which will finally take the space capsule out of the atmosphere of the huge planet. Clarke, however, does not allow the actual take-off to happen immediately but again resorts to a special "external intervention."

It turns out that a full five minutes has to elapse from the start of ignition to lift-off. The author absolutely needs this time, for two reasons: First it allows the fourth phase of the "attempt to make contact" in which one arrives closest to the level of noumenon in understanding the alien. Second, it creates the conditions to remove the burden of anthropomorphism from Falcon (a process which, however, cannot in the given circumstances be reflected in any way in the immediate action.) Without this unburdening, the final preponderance of the cyborg side of his dual identity would remain unexplained—a goal that is reached in the last chapter of the novella, which takes place back on Earth.

In the short span of time before the Kon-Tiki's ram-jet fires at full strength, there twice come physical touches between Falcon and the medusa. First, the medusa's tentacle "very gently rocked the Kon-Tiki," and soon after, "a large, heavy hand patted the balloon."

In both instances, Clarke deliberately shows the nature of the action of the "monstrous bag of gas" from the point of view of the human protagonist, so as to avoid the danger of the "omniscient story-teller" encroaching on the plane of noumenon of the alien. The expressions "very gently" and "a large, heavy hand patted" have meaning only for Falcon and not for the medusa. In this way the medusa's level of noumenon remains inviolate, but Clarke nevertheless succeeds in defining in the most accurate way up to now the nature of the creature's initiative, which is needed in order to arouse a particular reaction from the Falcon.

At first, Falcon tries not to pay any attention to the presence of the medusa's tentacle but when, on the second occasion, its inaggressive "patting" on the balloon becomes quite unmistakable, he can do nothing else but openly and for the first time allow the possibility that the bizarre Jovian creature does not have an unfriendly attitude towards him: "Of course, Brenner might be perfectly right. Perhaps it was just trying to be friendly."

And now follows the key moment of the whole novella. Thinking how to respond to this action of the medusa in which—burdened no longer by fear, just because of the very short time still remaining for him to spend in its vicinity—he perceives undoubted signs of a special wish for friendly contact, Falcon suddenly arrives in a blind alley. It becomes clear to him that anything he can do, even in this case, would be extremely inappropriate and laden with anthropomorphic limitations, albeit different ones to those produced by a negative emotional attitude.

"Maybe he should try to talk to it over the radio. Which should it be: 'Pretty pussy?' 'Down, Fido?' Or 'Take me to your leader?'"

The unambiguous irony in the second sentence simultaneously reveals two important things. On the one hand, Falcon has realized that the model of reciprocation offered him by parochial earthly experience—the possibility, that is, of a choice between a subordinate ("Down, Fido") or a superior or at least equal ("Take me to your leader") status for the alien being—becomes quite superfluous on a cosmic scale. In the context of Earth, "heterogeneity" is understood in a naively parochial way, as a mere difference in degree of homogeneous development. In this understanding, the possible essential and unbridgeable difference between truly heterogeneous (alien) beings are overlooked.

On the other hand, this very realization of the impossibility of establishing any kind of contact, even when the fear of attack is removed, finally tips the scales in the transformation of Falcon. The last phase of his withdrawal from the medusa, even though it started somewhat earlier and under different circumstances, has lost the underlying fear of the creature that was always present. Now there is an awareness of the impossibility of establishing contact while any anthropomorphizing ingredients remain in him.

Only in that light can one understand the real meaning of the sentence which Falcon murmurs at the very end of the chapter "Prime Directive," when the atmosphere of the largest planet of the Solar System, with all its threats and promises, is already far below him: "Some other time."

This "some other time" presupposes another, new, Howard Falcon, no longer burdened by anthropomorphic contradictions, who has at last attained his new, cyborg identity and has been liberated from the deficiency which stands in the way of the human race making contact with truly alien beings. Clarke does not in fact give much data about the advantages a cyborg has over a man in making contact with alien cosmic creatures, except that the cyborg would not be burdened by anthropomorphic contradictions. After all, anything like that would have gone beyond the scope offered by the framework of the story, since embarking on a closer analysis of Falcon's new identity would be as dangerous for the coherence of the work as an attempt at the noumenal conception and presentation of the Jovian medusa.

From the point of view of human participants in the work and the omniscient storyteller, the cyborg is as alien as the medusa, which means that Clarke had to be very careful as to what extent he could approach its noumenon. In the case of the cyborg, Clarke indeed did not need "external interventions" as with the medusa, because the basic story line could be developed right to the end without getting any closer to the noumenon of Falcon's new identity. The central meaning of the work was finally formed when it at last became clear to Falcon at the very end of the mission that any attempt at making contact with the medusa was destined to fail in advance, and would do so until he has transformed completely and thus freed himself from anthropomorphic restraints.

1.4   Conclusion

Consideration of "A Meeting with Medusa" has enabled us to arrive at the final conclusions of our investigation. This has primarily been centered on the problems linked with the possibilities of conceiving and presenting nonhuman characters from the point of view of the omniscient story-teller, in those science fiction works with the first contact theme that not parables.

We have seen that, in the novella we have been examining, contact between the earthly astronaut and the Jovian medusa could not have been made for two reasons.

On the one hand, such an event could not have taken place on Falcon's initiative, since anything like that would seriously impair the coherence of the psychological portrait of the pilot and of the work as a whole. On the other hand, the coherence of the work would have been seriously brought into question if contact had been achieved by the eventual initiative of the medusa, bearing in mind that, in this case, Clarke, in the role of omniscient storyteller, would have inevitably had to step onto the plane of motive, that is, of the noumenon of the heterogeneous entity, and would have thus violated a basic rule of the narrative.

But, although there has been no contact, Clarke, as omniscient storyteller, has nevertheless succeeded in getting closer to what we have designated as the level of noumenon in conceiving and presenting a nonhuman character. This was possible in the first place because Clarke resorted to what we have termed "external interventions" during the narrative procedure. These "interventions" represent special kinds of adjustments at the edge of plausibility, and they are used in the story in five particular places.

In the first two cases, Clarke from the very start stifles in Falcon a short-lived and hazily positive emotional attitude towards the "monstrous bag of gas"—an attitude which would in the long run have inevitably led to contact—in the way he abruptly introduces changes of situations. The first time involves the unexpectedly aggressive response by the medusa to the attack by the mantas, which strongly activates a mechanism of fear in Falcon. The second time Falcon's emotional wavering towards the medusa is replaced by an attitude which is fundamentally against making any contact is the sudden appearance of one of the gigantic creatures immediately above the Kon-Tiki.

The way in which the medusa and Falcon come closer to each other represents the third place at which an "external intervention" by Clarke saves the coherence of the novella. As we have seen, for this purpose, the author introduces a deficiency in the design of the space capsule with the obvious intent of showing that the above-mentioned convergence takes place accidentally—the only possible way, in the logic of things.

The fourth intervention is to leave to the medusa the special "initiative" in the four phases of the "attempt to make contact." Although the creature's actions are conceived in such a way that they do not at all suggest any possible reaching out by the omniscient storyteller to the plane of noumenon of the "monstrous bag of gas," they nevertheless display certain characteristics which dictate Falcon's irrevocable decision to withdraw, in the grip of a strong tide of fear.

It was just the necessity of such a response by Falcon that was for Clarke a reliable warranty that leaving the "initiative" to the medusa would not result in a contact that would seriously bring the coherence of the work into question. From the very beginning, in fact, Clarke created the psychological portrait of his hero in such a way as accord with this role.

In the process of examining to what extent it is possible, from the point of view of the "omniscient story-teller," to come closer to the noumenon of a nonhuman character, Clarke establishes that it does not suit him to have a character who is indubitably a man (that is, an anthropomorph) as the second participant in a "first contact" situation. For, as is convincingly demonstrated by the example of Dr. Brenner, who has no plan of action at the crucial moment, the necessity of withdrawal in the face of the medusa's "initiative" would not then have been warranted, an act which is the only thing that can save the coherence of the work. Only a special kind of transitional form suited this purpose, a form possessing exaggerated anthropomorphic characteristics, but also suggestions of a new, nonanthropomorphic status.

The particular half-cyborg nature of Howard Falcon thus represents a necessity that dictates the development of the central Jupiter episode in "A Meeting with Medusa." In addition to the considerably greater narrative space devoted to it, another factor that increases its importance is that some particular points from the episodes which take place on Earth can only be understood in the light of the events that unfold on Jupiter.

In this sense, the best example is the central symbol itself—that of the medusa—which only becomes functional at the very edge of plausibility, that is, when it is established that Clarke needed not only a cyborg but that kind of cyborg in whose consciousness a negative symbolic bridge could be established between the complex moment of transition from human to cyborg identity and some of the phenomenal, explicit features of the heterogeneous entity (form).

The psychological makeup of the main character is especially expressed in the fifth and last "external intervention" by the author, when the closest approach to the level of the noumenon of the heterogeneous protagonist is achieved from the point of view of the "omniscient story-teller." To make sure of an opportunity for realizing the "most intimate" degree of encounter between the earthly astronaut and the medusa, in circumstances where there is no longer any "danger" of actual contact, because the decision to withdraw has already been made, Clarke again manipulates the design of the Kon-Tiki.

This time, the "external intervention" involves the particular way the capsule engines work—they need a whole five minutes to reach full power. It is the very shortness of this interval, as well as Howard Falcon's perception that he is still not mature enough for contact, that each of his interpretations of the medusa's initiative will be based on anthropomorphization, until he has been taken over by his new, cyborg identity, which allows the author to ascribe one action to his nonhuman here—the "gentle rocking" and "patting" of the balloon—that in any other case would represent unwarranted reaching out for the plane of noumenon of the alien entity on the part of the omniscient storyteller.

Further than that—at least so it seems to us—one cannot go. The number of "external interventions" is even here on the very edge of what is permissible: their repeated introduction, with the aim of possible continuation of investigation of the possibilities of getting closer to the noumenon of the alien entity, would only have made an unconvincing construction out of a coherent and stable story. There is no doubt that Clarke was honestly sensitive to that intervention and did not allow himself further stretching of the bounds of probability.

But even in the framework within which he stopped, he succeeded in writing a work which, with regard to the noumenal conception and presentation of a nonhuman character, has almost no match in science fiction stories of "first contact."

Translated from the Serbian by John and Ružica White

2

Utopia in Arthur C. Clarke's *Childhood's End*

No Utopia can ever give satisfaction to everyone, all the time. As their material conditions improve, men raise their sights and become discontented with power and possessions that once would have seemed beyond their wildest dreams. And even when the external world has granted all it can, there still remain the searchings of the mind and the longings of the heart.
Arthur C. Clarke, *Childhood's End*

The novel *Childhood's End* resulted from a voluminous expansion of the novella "Guardian Angel" which originally appeared in two versions: a somewhat shorter, American one which appeared in the April 1950 issue of "Famous Fantastic Mysteries" and was edited by James Blish who condensed it and made minor alternations, and Clarke's original version—which was published in the Winter 1950 issue of the British journal "New Worlds". This latter one was subsequently used as the basis for the first of three parts of the future book.

The book being discussed here takes a special place in Clarke's SF writings, among other things because it presents the most complete axiology of the author's view of the world through a highly indicative sample of scientific Utopia. One of the advantages of treating this theme in *Childhood's End* is witnessed in the fact that it does not hold a key position in the structure of the plot but rather it occurs in a broader reference where conditions are amenable for studying it from external and internal perspectives.

"Utopia in Arthur C. Clarke's *Childhood's End*." Written in 1975. Originally published in Serbian in 1975 in the monthly magazine "Delo", 11–12 / 1975, 1617–1625, Belgrade, Serbia First published in English in "Foundation" #124, Science Fiction Foundation, Harold Wood, Essex, UK, August 2016, 85–91.

© Springer International Publishing AG, part of Springer Nature 2018
Z. Živković, *First Contact and Time Travel*, Science and Fiction,
https://doi.org/10.1007/978-3-319-90551-8_2

In placing the scientific Utopia in the coordinate of a cosmic history of the human race and not in an earthly one, Clarke found himself obliged to re-examine the plausibility of the function on which it is founded as well as the worthiness of the goals which it supports. Indeed, this re-examination did not essentially belittle science as a key factor on a specific level of the development of civilization, but it did point to certain general inadequacies in the Utopia which is founded on it—with regard to a much more relevant and general system of values than the ephemeral ideals of "the childhood" of mankind.

The scientific Utopia depicted in this work of the British author has a significant feature. It does not represent the fruits of human zeal, but rather comes as the result of external intervention by non-Earthly beings, whose degree of scientific advancement is incomparably higher than Man's. The motives of these "altruistic" deeds of the Overlords are not directly pertinent to our deliberation; furthermore, the human actors in the novel Childhood's End did not manage to grasp everything by the end of the novel, when it became clear that in the plans of the newcomers from cosmos Utopia was only a temporary and secondary phase whose background was devoid of only altruistic motives in the stricter sense of the word.

Clarke cites three conditions which enabled the Overlords to fundamentally change Man's world in a mere fifty years: "a clearly-determined goal", "a knowledge of social engineering" and "power". From the description of a subsequent realization of "the new world", however, it becomes clear that the first two conditions actually represent only prerequisites for the creation of a Utopia, while the focal point is exclusively found in "power". Clarke understands this term to mean the appropriate volume of scientific knowledge required to set up a positive form of control over the planet on which Man resides.

Just as in all Utopias, this control of Man's world is aimed at creating conditions under which every individual would be free of the obligations which hinder his creative activity. In *Childhood's End*, these conditions are treated in somewhat greater detail on two occasions, in chapters six and ten.

The emancipation of the individual-creator takes place on several levels, beginning with direct labour production all the way to professions in the world of entertainment, such as certain fields of sport. The Overlords first of all enabled the complete automatic production of basic consumption commodities, which completely eradicated the struggle for bare subsistence characteristic for all earlier periods. "The average working week was now twenty hours—but those twenty hours were no sinecure. There was little work left of a routine, mechanical nature. Men's minds were too valuable to waste on tasks that a few thousand transistors, some photoelectric cells, and a cubic metre of printed circuits could perform."

Idleness, which resulted from having a great deal of leisure at one's disposal, permitted everyone to devote themselves to more thorough and long-term education. Parallelly with increasing the general level of education, there was a final break-off with certain traditional mistaken notions of a spiritual nature which had burdened mankind, even when there was no real basis for this. Thanks to a device obtained from the Overlords, humans gained a direct insight into their own history, and into the period of the founding of all the more important world religions, which was sufficient to have them finally disappear. "Humanity had lost its ancient gods: now it was old enough to have no need for new ones."

The conditions for ideal material prosperity, or rather a very high standard of living, led to a dwindling of all ideological disagreements and to the disbanding of the standing armies. These global changes, as well as an entire range of smaller actions taken by the Overlords, led to a chain reaction of deep-rooted improvements in the general situation of many areas of secondary significance. The disappearance of state borders created "One World" from Earth and began from the ground up to do away with all race prejudices. The criminal had practically disappeared as the Overlords had the means for almost unlimited monitoring. Mankind had become exceptionally mobile, "and there was nowhere on the planet where science and technology could not provide one with a comfortable home, if one wanted it badly enough."

Some progress was made without the assistance of the Overlords. With the discovery of a completely safe oral contraceptive, as well as a reliable method for establishing paternity, the human race had finally rid itself of the last vestiges of puritan morals. Finally, the majority of people gained the opportunity to spend a good part of their time on various sports and entertainment in general, so that the whole planet slowly began to look like "a big playground".

In precisely this state we see the first cracks in the structure of the scientific Utopia—but cracks only for mankind and not for the Overlords, who never saw the Utopia as being the final goal but rather only the means. Although people had finally acquired irreproachable conditions for manifesting their creative potential, unhindered by the many restraints of the old world, idleness as a creative conditio sine qua non began to slowly transform into its negative correlate—boredom.

The course of this regression was reflected on a number of levels, but basically it had a uniform cause. The appearance of the Overlords and their uninterrupted presence had a very inhibiting and de-stimulating effect on Man's fundamental creative agent—curiosity. There was no longer any sense in wasting the whole lifetime on solving the mysteries of those scientific, artistic and philosophical issues which the Overlords, had perhaps discovered long ago.

This lack of enterprise became most evident in the field of art. "The end of strife and conflict of all kinds had also meant the virtual end of creative art. There were myriads of performers, amateur and professional, yet there had been no really outstanding new works of literature, music, painting or sculpture for generation. The world was still living on the glories of a past that could never return."

Stagnation in the field of science was partially hidden because there was an unprecedented boom in the so-called "descriptive disciplines" where facts were only collected and collated—so that almost no one even noticed the lack of theoreticians who would organize and link up these facts into a system. "Profounder things had also passed. It was a completely secular age."

Indeed, the race which had suddenly been guaranteed unlimited freedom and had been presented with inexhaustible sources of various kinds of entertainment—which "by the standards of earlier ages, it was Utopia"—had been so immersed in "the satisfaction of the present" that the anxious question of rare philosophers: "*Where do we go from here?*" did not reach them.

While this fundamental issue, just as in all preceding periods, remained on a purely academic level, the cracks in the scientific Utopia of the Overlords began to evoke suspicion among spiritually-minded people where they were most obvious: in the arts. The fact that stagnation had already turned into decadence in this field incited new debates on the motives and policies of the newcomers from space. "Was it possible that despite all their enormous intelligence the Overlords did not really understand mankind, and were making a terrible mistake from the best of motives? Suppose in their altruistic passion for justice and order, they had determined to reform the world, but had not realized that they were destroying the soul of Man?"

This first explicit criticism of the scientific Utopia led to the formation of a new Utopia which, indeed, was also founded on scientific grounds but whose ultimate purpose was not material prosperity but rather the return of the lost creative potential of people, who had increasingly turned into "passive sponges—absorbing, but never creating."

This new artistic Utopia grew at the site where the scientific Utopia began to lose its initiative and to close the spiritual horizons of Man. The artistic Utopia acquired its direct embodiment in the founding of a colony called "New Athens". The colony originated as the result of complex and voluminous plans in the field of social engineering, which served as the groundwork for reliably-defined optimal measures for the size of this community, its population composition, model of social order, as well as long term goals.

Still, regardless of this scientific guarantee, the founding of New Athens was awaited with a certain amount of skepticism for two reasons. In certain sense, the thus-conceived colony represented a challenge to the policy of the Overlords, who had never hindered the artistic ambitions of the people but nor did they encourage them. However, just as in many preceding cases, the newcomers from space did not react at all, remaining completely indifferent to all the activities of the Earth people which did not imperil the general welfare.

The second reason which partially generated suspicion in terms of the tenability of New Athens was founded on experience from earlier periods. "Yet even in the past, long before any real knowledge of social dynamics had existed, there had been many communities devoted to special religious or philosophical ends. It was true that their mortality rate had been high, but some had survived."

The ideal which was to have been embodied in New Athens was almost without precedent in the past. The basic concept of the founders of this utopian community was "to build up an independent, stable cultural group with its own artistic traditions". The pre-condition for these traditions consisted of providing a high concentration of world artists ("...nothing is more stimulating than the conflict of minds with similar interests"), who should achieve the optimum of creative utilization of idleness. "Everybody on this island", says one of the managers of New Athens, "has one ambition, which may be summed up very simply. It is to do *something*, however small it may be, better than anyone else."

It is obvious that the value of this work simultaneously defines the value of the artistic Utopia itself. The creative endeavours of the residents of New Athens were, first of all, concentrated on discovering original forms of expression, in the traditional as well as in the new artistic areas. It is, however, symptomatic that this aspiration toward originality, as an affirmation of creativity, was mostly reduced to a number of experiments reasonably described on one occasion as being "aggressively modernistic". As the number of experimental possibilities in the context of known artistic domains of expression was finite, there were rapid premonitions as to the final horizons of all fields of art.

Indeed, this fact did not threaten the creative potentials of the inhabitants of New Athens as generations were needed to finalize the already-initiated experiments. Still, the awareness of the existence of the final borderlines of art had a significant impact on the discreet occurrence of doubt as to the general value of this form of Man's spiritual expression or rather in its importance outside the narrowly local coordinates of Earth—coordinates which now had their incomparably broader correlate in the cosmic perspective of the development of mankind, a constant reminder of this being the presence of the Overlords.

In this situation, it was highly interesting but in a certain sense irrelevant to hear the opinion of the newcomers from space on the general value of art and, in the final analysis, on the usefulness of the Utopia called New Athens. An opportunity for this confirmation was shown during the visit of one of the Overlords—a visit which was supposedly intended to analyze the way of life and the goals of the colony but whose real motives were of a completely different nature. "There were some on the island who welcomed this visit as a chance of settling one of the minor problems of Overlord psychology—their attitude toward art. Did they regard it as a *childish aberration of the human race*? (*underlined* by Z. Ž.)"

The viewpoint of the newcomers from space concerning the value of the artistic expression of the Earth people was indeed difficult to define due to their reluctance to put forward any opinions which could even remotely suggest the final ends of their "altruistic" engagement with mankind. Still, there were two occasions on which somewhat more could be gleaned about this viewpoint.

In the first case, the conclusion was drawn indirectly, on the basis of the reaction of an Overlord when viewing a theatre performance. He actually reacted adequately and timely, but a certain doubt still remained. "He might himself be putting on a superb act, following the performance by logic alone and with his own strange emotions completely untouched, as an anthropologist might take part in some primitive rite."

On the second occasion, a concrete question was posed to the Overlord connected with the traditional dichotomy of the culture of mankind—the dichotomy between art and science—which concealed an intention to discover if, from the perspective of the Overlord, all artists actually represent abnormal individuals embodying "the childish aberration of the human race". The Overlord avoided giving a direct reply, making use of an ambiguous syllogism: "So if all artists are abnormal, and all men are artists, we have an interesting syllogism…"

The residents of the utopian commune could only guess at the real meaning of this syllogism, which simultaneously implied a judgment on the value of New Athens. In this respect, the readers of *Childhood's End* are in a somewhat more favourable position. They have the opportunity of attending the submission of the report by the Overlord who carried out the inspection of New Athens—a report which introduces directly, for the first time, a cosmic perspective in the process of assessing Man's attempts to safeguard the artistic form of expression in his creative spirit, something which had never been endangered in the past—at least not in Earthly frameworks. Tantalterresco said that no action should be taken in connection with the colony. Also, "It is an interesting experiment, but cannot in any way affect the future."

Now it is completely certain that the axiological judgment, condensed in the adjective "interesting", was really pronounced from the standpoint of "an anthropologist taking part in some primitive rite." From the perspective of the future, or the manifold cosmic usefulness of the human race, the utopian experiment called New Athens, which viewed the highest creative values of mankind in artistic expression, actually represents only "a childish aberration of the human race".

However, from this standpoint, any other type of Utopia—which represents the end and not only the means for universal cosmic development—is equally ephemeral and has no real impact on the future. Its lack of value follows from the static, non-progressive character of the mythical theme of "Paradise Regained" or the "Golden Age" which lies in the basis of any Utopia.

In *Childhood's End*, only the Overlords possess an awareness of transcience and instability of Utopia. At the end of the second part of the book, which quite intentionally carries the title "The Golden Age", the Earth Supervisor, Karellen, brilliantly summarizes the whole tragedy of this myth. "They would never know how lucky they had been. For a lifetime, mankind had achieved as much happiness as any race can ever know. It had been the Golden Age. But gold was also the colour of sunset, of autumn: and only Karellen's ears could catch the first wailings of the winter storms. And only Karellen knew with what inexorable swiftness the Golden Age was rushing to its close."

The easily overlooked issue of the last philosophers again pierce to the forefront in all its monumentality: "*Where do we go from here?*"

The cosmic dimension of the development of intelligent races—according to Clarke's concept—does not recognize Utopia. Those who have cast their lot with the Golden Age, regardless of whether this is founded on science, art or a third element, lose all importance in the order of the universe, turning into being which inexorably sink to stagnation and decadence. In order to reach higher levels of all-cosmic evolution, Utopia should be accepted only as a means and not as a final meeting of goals. On the other side of all Utopias, petrified in the ephemeral ideals from the period of mankind's "childhood", there are new dimensions of existence. The road to them sometimes stands in opposition to any altruism or the final end of each Utopia: self-satisfied prosperity.

Of course, nature always creates a prodigal abundance which permits for the prevalence of those races which could not muster up courage to confront this challenge and surpass the level of Utopia. "They have turned back while there was still time, avoiding both the danger and achievement. Their worlds had become Elysian islands of effortless content, playing no further part in the story of the Universe."

Translated from the Serbian by Irene Mirković

3

Chronomotion

What do we know for certain about movement through time?

Not much, actually. Only that, under normal circumstances, we advance at a steady pace of twenty-four hours per day in the direction from the past toward the future.

These normal circumstances can be violated, at least principally speaking, in two cases: if we alter the pace or the direction. The pace can be changed even in reality, and the direction, judging by all things, can be changed only in SF.

Let us turn our attention first to the pace, to speed.

It is possible to reach the future faster in two basic ways: suspended animation and (actual) fast movement.

Hibernation slows down physiological processes in the organism, including aging. It is possible to resuscitate a frozen person after ten years, for example, but they will have aged biologically only one year. The price for this quick arrival in the future is the loss of the time "slept through". That time, it seems, passes without dreams. It is no different from death, actually, except that resurrection is included in the price of the journey. Or it is, at the very least, available by choice. Anyway, it is a matter of choice. If something does not go wrong, of course. There is no absolute guarantee. But it is comforting that, if indeed something does backfire, at least the person will never know about it. . .

"Chronomotion." Written in 1995. Originally published in Serbian in 1995 in *Eseji o naučnoj fantastici/ Essays on Science Fiction*, Biblioteka XX vek, Belgrade, Serbia. First published in English in "Two Essays on Time Travel", "Foundation" #127, Science Fiction Foundation, Harold Wood, Essex, UK, August 2017, 77–83.

In terms of moving at high speeds (the closer to the speed of light the better: the effect of time dilation) the same result is achieved—one arrives far in the future with only the brief passage of local time. The advantage of this way is that one actively lives in local time. In your "local cosmos" (on a spaceship, for example) everything happens quite normally: The usual waking state exists, as does the usual sleep cycle (this time with dreams, including the bad ones as well, if you are already inclined to them or if you have a bad conscience). If you could observe the "external" world, it would seem to you that events go on there at incredible speed. And in reverse, to those "outside" it would seem that everything happening in your world would play out inconceivably slowly. But this spying on one another is not possible. Two temporal streams are irreconcilably split. In reality, that is.

Does the accelerated pace of movement through time offer any literary consolation? Certainly. Human drama.

Above all, melodrama. Are there any more exciting melodramatic situations than the meeting of close relatives in completely mistaken time phases?! For example: Just after giving birth to her daughter, a young mother sets off on a time dilation journey which will last only one year in her local time, but then, when she returns, she will encounter the world from whence she came which is twenty years older, so that she will meet her daughter who is now the same age as she. (Ursula Le Guin, "Semley's Necklace".) In a literary sense, the idea seems steadily attractive, although the terrain here has already been thoroughly investigated, because the factor of estrangement is relatively simple.

The extreme cases of accelerated movement through time are those which occur in (only genre allowable) points of reaching and exceeding the speed of light. Would time then stop completely, or might it even begin to go backward? Would one here achieve regression (or 'progression', depending on one's point of view) from old age to youth, and even further, to birth? (Dan Simmons, *Hyperion*.) What comes before (that is, in this backwards view, after) birth: death? Could it be considered that a man is dead before he is born? The estrangement is much less restricted here, but it is not certain that the human drama also grows proportionally. On the contrary. Everything in good measure. Extremes are an uncertain means of support.

The genre also allows the possibility that temporal streams with different rates of movement into the future do not remain separate after all, but that they come into contact, interweave. What happens when a slow and a fast temporal stream meet? What kind of influence can they have on one another? Would the slower one be more authoritative than the fast one? (Roger Zelazny, "The Great Slow Kings".) Or, perhaps, is the opposite true? (Frank Herbert, *The Heretics of Dune*.) The idea is not void of genre excitement, but it can seem to be a bit stretched, unconvincing. At least as fertile ground for high quality human drama.

Finally, what is the last moment at which, in one of two accelerated ways, one can arrive in the future? Is there, in and of itself, such a moment? The end of time? The end of the cosmos? Does the idea of the end of time make any sense whatsoever? What if time is not linear but cyclical? What if, instead of advancing, the only place where it is appropriate to talk about the end, we have an eternal returning, a melding of the end into a new beginning? This is stepping into the area of ultimate questions, of cosmology, and here human drama, after all, becomes secondary.

Or does it? It is not completely certain. Reading the novels of Olaf Stapledon (*Last and First Men*, *Star Maker*) or some of Arthur Clark's stories ("Transcience")—written under the influence of Olaf Stapledon—most certainly does not leave the reader indifferent. But what is exciting here is not so much the human drama as it is the vision which reaches immeasurably beyond human boundaries. The question remains open as to whether such visions can possess literary value and also human drama.

So much for speed. Let us look at what happens with change in direction. Specifically, movement from the future into the past.

One might say, at first glance, that in reality there is nothing to support such an idea. No time machines, at least as far as we know, have been noticed arriving from the future. Not now, and not at any earlier time.

This, to be fair, still does not mean anything. Perhaps the world is teeming with such time machines but we do not notice them, because they either directly or indirectly are not making their presence known, that is, they are not influencing the past. Why don't they? It is possible to come up with three principle reasons: They don't want to, they don't dare to, or they are unable to.

If they dare to and are able to, but still do not want to, then only one cause for that can be discerned. They are behaving as impartial observers who are avoiding getting involved in any possible way with the object of their observation. Those would, actually, be the ideal historians: They have a chance to directly study the past which they have absolutely no desire to desecrate by interfering with it. (Connie Willis, *Doomsday Book*) Here there is plenty of dramatic tension, but it is somewhat simplified: everything, actually, is reduced to whether it is possible to retain complete indifference toward the time being observed. The desire to take action in it can be truly powerful. It is fertile ground for an exciting story, but it basically remains summarily simple.

If they want to and are able to, but do not dare to, then it can be supposed that there is some sort of injunction keeping them from it. The purpose of the prohibition could be the protection of the future which can be unexpectedly and irreparably disturbed if the fine weave of the past is changed even in the slightest. Even the most harmless causes in this time can have truly heavy consequences in a later one. (The "Butterfly Syndrome"; chaos theory.) In order to avoid that, mechanisms or institutions are introduced to more or less effectively protect the past. (Poul Anderson, "Time Patrol", or Isaac Asimov, *The End of Eternity.*) This is also not void of human drama, but in terms of the genre it is a bit unsophisticated and old-fashioned.

Finally, if they want to and they dare to but are unable to, then most likely there is a paradox at hand which, so it seems, is inescapably accompanied by attempts to change the past. The most familiar among them is the following.

A man goes back into the past and either accidentally or intentionally causes the death of one of his parents before he himself is conceived, thus thwarting his own birth. But if he was never born, then he could not return into the past and prevent his own birth by getting rid of one of his parents, so he was born after all and returned into the past where he thwarted his own conception, which means that he was never really born... And so forth. *Reductio ad absurdum.*

This nonsensical mixing of cause and effect remains, judging by everything, as the final prevention of altering the past from the future, but it does not stop the writing of works for which time paradoxes are no kind of barrier. Those paradoxes are simply not taken into consideration here. What improbable results can then be achieved! (Robert Heinlein, "All You Zombies...".) The problem, however, lies in the fact that here there is no real drama: it is mainly just being clever and witty in the genre. Who will conceive of a more perplexing thing than the others. A mere *Gedankenexperiment*, a thought experiment.

In the end, what remains?

Clearly, the challenge in the genre is greater if changes are accepted not in the speed but in the direction of movement through time. Ideally, a means of influencing things in the past should be found while avoiding the paradox trap which that influence implies. Is something like that at all possible? It is not, so it seems, if there is a presupposition that there is only one primary stream of time.

If, on the other hand, one thinks that there are several streams (an infinite number), then the aforementioned paradox can be elegantly surpassed. In the place where the past is changed, a branching out occurs: Along one branch—the one from which a time-traveler sets out to intermingle "backward" in time—the change has no influence (that stream has already played out the way it did), but another, new branch appears which is completely dependent upon the change. Transforming the past does not change, therefore, the future, but rather creates a new stream of time which exists in parallel. (Ursula Le Guin, "A Fisherman of the Inland Sea".)

This is a case of a first-class idea in the genre about a specific kind of chrono-tree with literally infinite branchings and forkings. The axis of the powerful dramatic tension here is created by the question of whether events in one stream can affect events in another, that is, is it possible to move from one stream into another? Likewise, what is the relationship between the different versions of the same character in the different streams of time? Obviously, this understanding of the chronomotion theme is also not bereft of the danger of paradox, but that does not have to be crucial in a literary sense. Science fiction is not an experiment in theoretical physics, but a work of fiction. And works of prose allow paradoxes. In moderate doses, it goes without saying, and especially if they are well conceived.

Translated from the Serbian by Randall A. Major

4

The Labyrinth Theme in Science Fiction

In traditional literature, as well as in other likewise traditional forms of artistic expression, the theme of the labyrinth appears exclusively as a spatial phenomenon. In that context, it possesses three ontological properties which differentiate it from other similar themes, for example the themes of wandering, searching, that is, traveling in general.

The first property would be movement in a defined, enclosed space, which essentially can have only two directions: toward the exit or back toward the entrance. The second property would be the existence of a dead-end street, a no way out of the basic peripatetic factors of the theme in question. Finally, the third property of the labyrinth theme is the facilitation of essential change in the protagonists during the story's plot, which is, as literary theory teaches us, an essential condition of good fiction: Protagonists are not the same at the entrance and at the exit of the labyrinth.

Within the genre of science fiction, the theme we are discussing has undergone a fundamental modification which caused certain alterations to its ontological properties. In SF texts, the labyrinth theme is no longer a *locomotional* (spatial) but *chronomotional* (temporal) phenomenon. Movement from the entrance to the exit here, however, is also limited by a closed system, which can also have only two directions: from the past toward the future, or from the future toward the past. Time has no third dimension.

"The Labyrinth Theme in Science Fiction." Written in 1980. Originally published in Serbian in 1981 in the monthly magazine "Delo", 1–2 / 1981, 150–154, Belgrade, Serbia. First published in English in "Two Essays on Time Travel", "Foundation" #127, Science Fiction Foundation, Harold Wood, Essex, UK, August 2017, 77–83.

Yet, while both directions are possible in the model of the spatial variant of the labyrinth theme—toward the exit or back toward the entrance—and they are basically of equal impact, when discussing the temporal variant, there is one significant difference which occurs in the existence of the dead-end street. While in the spatial labyrinth it is possible to reach a dead-end street whether one is moving toward the exit or toward the entrance, one can reach a no-way-out situation in the temporal labyrinth exclusively while moving from the future toward the past.

In that direction, the dead-end street appears in the form of certain logical and causal paradoxes, retaining, however, its basic function as a peripatetic factor. Finally, when discussing the third ontological property of the labyrinth theme, in the temporal variant as well there is a transformation of the protagonist, but in this case it is generally much more drastic than in the spatial variant: in the first case, the change in the characters occurs exclusively in terms of their *weltanschauung*, while in the latter they can change their entire generic identity, mostly due to the effect of logical and causal paradoxes related to the specific nature of the dead-end street within the temporal variant.

It is obvious, therefore, that the focus of the *chronomotional* type of the labyrinth theme lies in the factor of the dead-end street, which basically appears only in one morphological type, with a wide range of variations. All SF works which utilize the concept of movement from the future into the past are faced with the obligation to take the following paradox into consideration. A protagonist who lives in time A returns to the past in time B, where he does something which will result in his disappearance from the future; this disappearance is motivated by linear causality: the hero, for example, kills one of his ancestors before he is able to leave his progeny behind, and thus the hero's very own birth is thwarted. The chain of causation, however, does not end there. If the protagonist is not born, meaning that he did not exist in time A, then he could not have possibly returned to time B and kill his ancestor there; and in that case, the ancestor did manage to leave his progeny behind, and thus the protagonist is inevitably born in the future!

Thus, we have a paradox, a dead-end street. If it is presupposed that he returned to the past and killed his ancestor there, the protagonist could only be born if he was never born! *Contradictio in adiecto.* We have been led there by the disciplined application of linear causality, which is unmistakably in force in the case of the spatial labyrinth: Namely, it has long been known that the absolutely certain way of getting out of every closed system of this type, regardless of how complicated it is—all one has to do is follow one wall, starting at the entrance, and sooner or later one has to find the exit.

Here, cause and effect act in a linear fashion. There is, however, one significant moment which is usually overlooked because it is not so obvious: We said—*sooner or later*. In other words, between cause and effect there is always a given temporal gap, only it is always true that the cause (in terms of linear causality) *comes before* the effect. One cannot, namely, first exit from the labyrinth and only then start following one wall beginning at the entrance. Ultimately, cause and effect are, in every practical sense, simultaneous (when you press the switch the light practically—although not really—goes on in the room).

However, can such a sequence of cause and effect remain in force when discussing *chronomotion* from the future into the past? It is clear that in this case the linear flow of time does not work, in the sense that—already depending on how we look at things—in certain cases the future precedes the past. Concretely, if the act of murdering an ancestor is understood as a consequence of the protagonist's setting off from the future into the past then, from the perspective of some sort of absolute time, it would seem that here the consequence (in the past) preceded the cause (in the future), which is in opposition to the fundamental principles of linear causality.

The objection could be made, however, that absolute time is not authoritative here, but rather the individual time of the protagonist. At the moment of the ancestor's murder, regardless of the fact that it takes place in the past, he is older than at the moment when, although it happens in the future, he set off on his time travel. In that case, the cause would precede the effect after all and linear causality would be preserved.

Accepting the authority of the individual time of the protagonist in defining the chronological order of cause and effect confronts us, however, with another difficulty. The individual time of the hero could serve as a valid measure for the establishment of the order of cause and effect in *chronomotive* cases, if it were not for the paradox of the multiplication of the protagonist. Let us imagine the following situation: The protagonist is located in a given space at moment A. Half an hour later, he gets in a time machine and returns thirty minutes into the past, precisely at moment A. At that instant, now, there are two protagonists at the same place, who are completely identical in every other way except one: one of them is half an hour older than the other, meaning that their individual times are thirty minutes different. Nothing is preventing such replicas of the first version of the protagonist from appearing in unlimited numbers, and the difference between their individual times could increase to any value within the normal human lifespan. In such a situation, since ultimately one and the same protagonist is in question, it is no longer clear which of his differing individual times should be chosen as authoritative.

Linear causality, thus, is inapplicable to the situation that results from the *chronomotive* premises about movement from the future toward the past. In other words, in the temporal labyrinth there is no wall that we could follow and thus certainly find the way from entrance to exit. In that case, how does one find one's way around in it? Is there some alternative linear causality which could eliminate the aforementioned paradoxes and remove the dead-end street?

Yet, is this question, perhaps, faulty from the outset? Perhaps science fiction has no ambition whatsoever to eliminate paradoxes and remove dead-end streets. Perhaps SF really wants to have them. Finally, regardless of the prefix "science", it is still fiction. Didn't one famous fantasy writer lucidly observe that sometimes the path to the goal is more important, the labyrinth more important than the exit, the paradox more important than a clear solution, and the dead-end street more important than the wide avenue? Because, if everything could be reduced to causality, we would have, to be honest, a mathematically perfect world, but it would be quite difficult to say that great art is one of its virtues.

Translated from the Serbian by Randall A. Major

5

Annotations 1

Science fiction has made two major contributions to the thematic treasury of the art of prose: time travel and first contact. Both these themes had already appeared in the opus of one of the founding fathers of the SF genre, Herbert George Wells: time travel in *The Time Machine* (1895) and first contact in *The War of the Worlds* (1898).

As for time travel, Wells brought out the main literary value of this theme: human drama. If it is intense enough, it can camouflage the inevitable paradoxes that pop up everywhere in a chronomotion story, threatening to ruin its delicate narrative coherence.

With regard to first contact Wells established a model for the non-human protagonist: invaders who come to Earth to conquer its human inhabitants. This model reflected our profound fear of the menacing vastness of space, potentially full of unknown threats. The idea of malevolent extraterrestrials was very present in the first half of the 20th century, in both SF literature and cinematography.

Only in the 1950s did there begin to appear, rather timidly, literary works in which the Others were more or less benevolent: missionaries rather than invaders. It took us even longer to realize that there was a third possibility—that the Others might be neither good nor bad, but indifferent.

When we imagine Others, in our SF works, as either invaders or missionaries, either good or bad, we always anthropomorphize them. We tailor them to our own measures. We project our own motivations on them. But it is only really safe to suppose that they are fundamentally different from us, entities with inconceivable motivations, far outside our anthropomorphic norms.

© Springer International Publishing AG, part of Springer Nature 2018
Z. Živković, *First Contact and Time Travel*, Science and Fiction,
https://doi.org/10.1007/978-3-319-90551-8_5

If this is so, then we have to face a crucial question: do we even have the mental ability to imagine genuinely heterogeneous entities? Others which would not be in the least anthropomorphic?

This question was the pivotal one in my MA thesis. I examined it through the SF opus of Arthur C. Clarke because it contained works that were exemplary, both for differentiating various types of anthropomorphism and for establishing how far one can go in imagining first contact between humans and a truly heterogeneous entity.

Many of Clarke's first contact novels and stories were not taken into account in my thesis. This was because they were not relevant to my investigation. For example, in Clarke's two major first contact novels—*2001: A Space Odyssey* (1968) and *Rendezvous with Rama* (1973)—there are no Others at all, only their artefacts. In his letter to me, written after he had read my essay on the first contact theme in his SF works, Clarke expressed surprise that I hadn't mentioned his story "Rescue Party". It is indeed an excellent first contact story, but again, not relevant to my study.

Clarke's novel *Childhood's End* (1953) is another fine example of his first contact works. I did not mention this either in my MA thesis, although it is one of the first books in the history of science fiction in which benevolent aliens appear. I was, however, interested in it because of another theme it dealt with: Utopia. Although much older than science fiction, the utopian theme was generally considered SF "property" in the 20th century.

To conclude, the essay section of this book contains my early works, mostly written about 40 years ago and covering two themes unique to science fiction: first contact and time travel. I find the fact that they are still readable (and publishable) after nearly half a century rather flattering. But they are not my last word on these themes. I returned to them much later in life when I became a writer myself. Because there are things one can only say as a writer.

Sir Arthur C Clarke

'LESLIE'S HOUSE'   25 BARNES PLACE,   COLOMBO 7,   SRI LANKA.
PHONE: (941) 694255, 699757   FAX: (941) 698730.

Zoran Zivkovic   9 April 2001.
Bulevar Mihajla Pupina 10E/162
11070 Novi Beograd
Jugoslavija

Dear Zoran,

Thank you for the first *Contact* articles which I read with bemused interest. I must confess that I'd completely forgotten *Crusade* and had to look it up, and had only a few memories of *Medusa*. I am surprised though, that you didn't mention, perhaps my most important "first Contact" story *Rescue Party* with its extremely chauvinistic last line.

It's now some years since I've written any fiction - far too busy dealing with visitors, correspondence and email! See attached schedule.

Anyway, thank you for your interest in my work.

All good wishes,

art clarke

Enc: Schedule

An Arthur C. Clarke's private letter to Zoran Živković

Dear Editors:

As usual, the February 2002 *Interzone* inspired a few thoughts:

1. Concerning J. R. R. Tolkien – I have been trying for years to find out if the poem by Steve Connolly ("God bless the Squire," etc.) was serious or a parody. The first time I met Tolkien, I didn't realize who he was till years later – I was too busy talking to his companion, one C. S. Lewis. Next time it was at a literary luncheon. He sat next to me and pointed to the rather small man at the end of the table, his publisher. "That's where I got the idea for *the hobbits*," he whispered.

2. I much appreciated Zoran Zivkovic's remarks about me and am flattered that I helped him get his MA. As far as I know the only other university thesis on my work was written by a Russian lady. (How about it, England?) I think that someone who translates 16 of my books deserves a special medal.

3. I must confess that my Thogism ("Ansible Link," *IZ* 176, p46) was home-brewed – I had it around for years and was happy to find a home for it.

4. I enjoyed Tim Robins's piece on *Star Trek* – and enclose a piece of history that wasn't mentioned! *[Photocopy of a letter from Gene Roddenberry, acknowledging Arthur C. Clarke's influence on the TV series – Ed.]* I was certainly proud to have saved *Star Trek* – but am even prouder of something that may be slightly more important. I've recently learned that Dr Werner von Braun used my book *The Exploration of Space* to persuade JFK to go to the Moon.

Keep up the good work!

Sir Arthur C. Clarke

Colombo, Sri Lanka

An Arthur C. Clarke letter published in the May 2002 issue (p. 4) of Interzone

Part II

Fiction

6

The Bookshop

The fog, as usual, set in swiftly.

Only a few minutes had passed since the last time I'd raised my eyes from the computer screen and looked out of the bookshop's large display window. In the early twilight I had been able to see buildings on the other side of the river quite clearly, speckled with the first evening lights. Now everything had suddenly disappeared in the thick greyness; not only the opposite bank but also the long row of horse chestnut trees extending along the quay on this side of the river, just a few steps away. Although this transformation had taken place almost every evening since the middle of autumn, it never ceased to fascinate me. One moment the world was there, real, visible, tangible; then, in what seemed like the twinkling of an eye it would magically dissolve in the humid breath of the river spirits.

I could have closed the bookshop and gone home. For days no one had entered the shop after the fog rose. In autumn the river reversed its genial summer personality. When the weather was warm, the promenade under the horse chestnut trees was thronged till late in the evening. Then I would often stay open until midnight and sometimes even later, until the last customer had finally finished leafing through what I hoped would shortly be his book. The customer has always come first in this bookshop. But now I remained in the shop not only because the shop hours posted on the door obliged me to. I did not have a computer at home, and it seemed somehow inappropriate for me to write science fiction in the old-fashioned way, pen to paper.

"The Bookshop." Written in 2000. Originally published in Serbian in 2000 as "Knjižara" in *Nemogući susreti/Impossible Encounters*, Polaris, Belgrade, Serbia.

© Springer International Publishing AG, part of Springer Nature 2018
Z. Živković, *First Contact and Time Travel*, Science and Fiction,
https://doi.org/10.1007/978-3-319-90551-8_6

But tonight I was not to be allowed to return my attention to the screen. My eyes were still gazing, unfocussed, at the wall of mist on the other side of the window, when a figure took shape in front of the entrance, seeming to materialize out of nowhere. Its sudden appearance, unannounced by any footsteps on the pavement—unless, lost in thought, I had simply not heard them—made me start. Fog is apt to produce such eerie surprises, and I disliked it almost as much for that as for taking away my customers.

The man who came in was small and slight, with a short, sparse beard and wire-rimmed glasses. Although he appeared youthful, his grizzled sideburns and the silver streaks in his beard, particularly on his double chin, strongly suggested that he had passed the half-century mark. I have a good memory for faces, so one glance was enough to tell me that I had never seen him here before.

It must have been rather cold outside, for no sooner had the visitor entered the heated air of the bookshop than moisture condensed on his glasses, fogging them up completely. He stood by the door without moving, seeming to stare fixedly at me through large, empty eyes of unearthly blankness.

I pressed two keys at the same time, saving the text. This was not really necessary, as I had made no changes since the previous save, but that is what I always do, automatically, whenever there is about to be a break in work.

"Good evening," I said. "The fog is really thick tonight."

The man took off his glasses. He rummaged for a while through his long, green coat until he found a crumpled white handkerchief in an inside pocket and started to wipe his glasses. His movements were brisk and impatient, and left patches of condensation by the edges of the frame when he put them back on.

"This is a science fiction bookshop." It was somewhere between a question and a statement. There was something strange about the way he drew out his vowels, as if he were a foreigner who had learned the language well, but still hadn't quite mastered the proper accent.

"That's right," I replied with a smile, "*Polaris*. At your service. If it weren't for this terrible fog you wouldn't have to ask. There's a large neon sign above the entrance, but what good is it now? I paid a ton of money for it, but they forgot to tell me that it's completely useless in the fog. It would probably be better to turn it off. Drives customers away more than it attracts them. Even when you're right under it, it just looks like a bright, shiny rebus."

Still standing by the door, the visitor began to look around the shop. He slowly skimmed the shelves full of books and magazines, appearing somewhat bewildered, as though he had entered some amazing place, and not an ordinary bookshop at all. That is to say, maybe not exactly ordinary, since science fiction bookshops are a bit unusual, but they don't generally induce such bewilderment.

"I'm looking for a...work of science fiction," said the man, after his eyes had finally reached the counter with the cash register and computer, next to the display window, where I was sitting. His voice sounded hesitant, as though he had trouble choosing his words.

"Then you've come to the right address," I replied cordially. "We offer a wide selection of science fiction—new editions and secondhand. We really pride ourselves on them. We've got some truly old books. Real rarities you won't find anywhere else. And should we happen to be temporarily out of what you want, we can get it very quickly. In two or three days at most."

The visitor finally moved away from the door and headed towards the counter. He stopped uncertainly when he got close to me, as though not knowing what to do with himself. I got a sudden whiff of a fresh, outdoorsy smell. It immediately brought to mind newly mown grass. The man must use a deodorant based on plant extracts.

"The work I'm looking for is in this bookshop," he said. His tone had lost its previous uncertainty and become self-confident. Even more than that: he said it in a voice that would brook no objection. "And it's not old at all. Quite the contrary, it's just been written."

"In that case," I replied, "it must be here." I got up from my chair and headed towards the shelf where I kept the latest editions. "Here you are."

Seven narrow rows contained some fifty books that had been published in the last several months. Science fiction was on the upswing again. This time last year those shelves had held barely fifteen volumes. I reached towards the middle shelf and pulled out a rather small book with a shiny cover.

"This is our most recent acquisition—*Impossible Encounters*. Might this be what you are looking for?"

The customer briefly examined the book in my hand, then shook his head. "No, that's not it."

"I suggest you have a look at the other books. These are all recent editions."

I left the visitor in front of the shelf and returned to the counter. People don't like you to hover round while they leaf through a book. It gives them an unpleasant feeling of being under surveillance.

My eyes dropped to the screen, with its tangle of words. The story I was writing was practically finished. All that was left was to read it once again and polish it up here and there. I would have had no trouble doing so in the solitude I'd expected until I closed the shop. Now that solitude had been interrupted, but I hoped the man would quickly find what he was looking for so that I could resume my concentration on the text. I could not, of course, work while he was there. Not knowing what else to do while I waited, I pushed the 'save' keys once more.

My fingers were still on the keyboard when the visitor came up to me again. At first I thought he'd found the book he wanted, but when I raised my eyes I saw that his hands were empty.

"It's not there," he said.

"You've already looked at everything?" I asked, unable to conceal a note of disbelief.

"Yes, there are only forty-eight books," he replied in an even tone. If he'd noticed the surprise in my voice, he did nothing to show it.

I gazed briefly at the man in front of me, and then at the shelf with new editions. "Why, yes," I said at last, "only forty-eight."

"Where else could I look?" he asked rather quickly.

"If it's a really new book, then that's the only place it could be. I don't keep them anywhere else. The other shelves contain older editions. Which book are you looking for? If you tell me the title, I can help you find it."

"Title?" The visitor squinted in dismay through his glasses, which were now dry. "I don't know the title."

"It doesn't matter," I hastened to assure him. This was by no means a rare occurrence. I encountered variously incomplete requests almost every day. "The writer's name will be enough. That will make it easy for us to find the book."

The man took his handkerchief out of his pocket once again and wiped the top of his forehead. He was clearly dressed too warmly for indoor temperatures, and beads of sweat had started to break out. I was assailed by another outdoor smell. Instead of mown grass it was some wildflower this time, but I couldn't determine which.

"I don't know the author's name." A look of unease crossed his face.

I sighed inwardly. Any chance of finishing work on my story that evening was receding. This was likely to take some time.

"Why don't you make yourself more comfortable," I suggested. "It's rather warm in here, and it may take us a while to find this work, with its unknown title and unknown author. You can leave your coat on the hook by the door."

The visitor shook his head briskly. "No, no. I can't take off my coat. I don't have much time. It's an urgent matter. I have to find the work as soon as possible. I can't go back without it. You don't understand..."

He said this very quickly, in one breath, and then suddenly stopped, as though for some reason he couldn't or didn't want to continue. A pleading look came into his eyes.

"I do understand," I replied after a short pause. "You want to find a specific work of science fiction and you are in a hurry. I certainly want to help you, but you have made only very scanty data available to me. All that I know is that it is some new work and that you didn't find it on the shelf over there. If you could tell me something more about it, I might recognize it. I read a lot, almost everything that comes out. Particularly new things. Could you at least give me some idea of what the work is about?"

A smile played on the man's lips. "That I can do, yes. Certainly. It is about my world."

We stood there several moments looking at each other without speaking. I was smiling too.

"Your world?" I repeated, breaking the silence first.

"Yes, but you on Earth know nothing about it. Or rather, nothing was known until recently. Until the work I am searching for was written. Our star doesn't even have a name here, just a number, although it is relatively close, less than eleven and a half light years away. But it's a small star, much less conspicuous than those around it, so there's nothing strange in it being anonymous."

I slowly nodded my head to indicate understanding, as if he were telling me something quite commonplace. So that was it. One more of those. Yet he hadn't the look of one. Quite the contrary. But appearances can be deceptive, as had been proved often enough. Clothes alone do not the eccentric make.

All kinds of oddballs visit my bookshop. They seem to be irresistibly drawn to it, and they constitute an ineluctable hazard of my chosen genre. I am most often visited by those who have had first-hand experience with extraterrestrials, and for some reason feel this is the right place to bare their souls. At first I entered into discussions with them, explaining that I class science fiction as imaginative prose. Their real-life experiences had no place in this category, for the very reason that they were real. As a rule, however, this distinction was too fine for them.

Then, in my naivete and inexperience I tried to talk them out of it. Why go to the inconvenience and expense of shooting across from the other side of the cosmos, only to subject some commonplace citizen in an isolated house to unusual lights or sounds? That was when I got into serious trouble. Not only did they turn a deaf ear to the reasons I cited, they resolutely interpreted my unwillingness to believe them as reliable confirmation that I, too, was part of the great conspiracy to hush up visits by extraterrestrials. That was the milder version. Several flying saucer fans accused me openly and rather peevishly of being an extraterrestrial myself.

There is no complete defense against such accusations. Indeed, how can anyone prove he is not an extraterrestrial to someone who can see antennae sprouting from his forehead? What arguments can ever shake the believer's blind conviction? But to me the primary difficulty stemmed from my profession. As the owner of a bookshop I could hardly draw distinctions among my customers based on their view of the world, so my hands were tied. Should I meet this type of person in some other context, I could solve the problem simply by raising my voice. A slightly sharper tone has a truly amazing effect on them. They fall silent at once and withdraw, often in embarrassment. But here, that would be out of the question. How would it look if a bookshop-owner yelled at those customers who just happened to take a somewhat unusual view of his ancestry?

And so I resorted to the last means still at my disposal. Whenever an eccentric like this one drops in, I listen to his story with utmost patience, regardless of how far-fetched it is, taking great care to speak as little as possible. My most frequent reaction is to nod or shake my head from time to time, as befits the situation, to demonstrate that I am carefully following the story. This technique has often proved useful. First of all, the whole affair is concluded far more quickly than if one were to start a discussion; second, after baring his soul almost every single visitor of this kind ends up buying a book.

Over time this proved adequate compensation for approximately a quarter hour of my attention. I could almost have made this part of my price list: "The purchase of a book gives the buyer the right to squander fifteen minutes of the owner's time in any way he sees fit". At first my conscience bothered me a bit, feeling this partook of prostitution; then my business sense over-rode such improvident moral purism.

Furthermore, over time I came to see myself as a psychiatrist—a rather poorly paid psychiatrist, it's true, but at least there was never a shortage of patients. Quite the contrary. There were so many of them I could no longer rely on memory alone, and had had to buy a notebook in which to write down what each one of them bought, so they would not accidentally buy the same book twice. This, to be sure, didn't bother them in the least, since most of the books were never read—occasionally I even found them discarded next to a nearby trashcan—but for me this was a matter of professional attitude towards my work. Every customer deserves the best possible treatment, and the handicapped get a bonus to boot.

But never before had I encountered a case like this. This was the first time that an extraterrestrial had visited my bookshop! Perhaps I should have been jealous. Up till that moment the role had been reserved for myself. Granted, the situation hadn't changed essentially. It was just a matter of nuances. My basic strategy remained the same: don't question anything and encourage the speaker to tell his story without holding back.

"Eleven and a half light years," I said. "Why, that's really not so small. You had to travel quite a distance! It must have taken you a long time."

The man shook his head. "No time at all. It's hard even to call it travelling."

"I see. Did you spend the flight in hibernation, then? Is that why it seemed so short?"

"No, hibernation wasn't necessary."

"Oh. Then that means you must have a very fast spaceship. Judging by how quickly you got here, it must travel considerably faster than the speed of light."

He looked at me the way a teacher looks at a student who has blurted out an absurdity. "No spaceship can travel faster than the speed of light."

"Of course it can't," I said, hastening to correct myself. "How silly of me. I forgot that for a minute. Then how did you get here so fast? Excuse me for not being able to figure it out for myself—space travel is not one of my strong points."

"In the only way possible. Using the fifth force."

It's not easy to carry on a conversation like this. One must keep a straight face, and there is great temptation to poke fun. It's even harder to suppress the laughter that is ready to bubble to the surface. But through long experience I have become very skilled in self-control.

"The fifth force?" I repeated, expressing the mild surprise I felt appropriate.

"That's what we call it. You know about it, too, but haven't yet recognized it as a force, so you use another name. Actually, it has several names. One of them, for example, is imagination."

This time I didn't have to feign surprise. "Imagination?"

"Yes. Imagination, fantasy, daydreams, whatever you like. The ability to conceive of something that does not seem to exist." He indicated the shelves around us with a broad, sweeping gesture. "All these are the fruit of imagination, aren't they?"

I could only confirm that they were.

"And you are convinced that they are pure fantasy. You feel that there's no way the worlds of science fiction could ever be real. Isn't that right?"

"Well...yes..." I mumbled, finding myself in a spot. "I mean, for the most part... Although sometimes, of course, there might be certain coincidences... It's not out of the question... But very rarely..."

"Tell me," he said, putting a stop to my stammering, "how does a work of science fiction originate?"

I didn't reply at once. The conversation had taken a completely unexpected turn. Who would have thought that we'd wind up discussing the problem of literary creation? I have discussed many unusual subjects with the eccentrics who visit me, but never this.

"Well, I don't know exactly. My experience in this regard is quite limited. I have only written a few stories. I suppose the writer cogitates, and then an idea flashes in his mind and..."

"An idea flashes, yes! Do you know what actually happens at that moment—when, as you say, an 'idea flashes', seemingly out of nowhere?"

Of course I didn't know, so I shrugged my shoulders.

"The fifth force is activated!"

The pause that followed was deliberate, a dramatic effect calculated to ensure that the revelation would make the strongest possible impression on me. To demonstrate enlightenment, I nodded sagely.

"Unlike the four fundamental forces that exist on the level of the very simple, the fifth force appears solely on the level of the very complex. It can take effect throughout the cosmos, but in only a single class of locale: in centers of awareness of sufficiently developed species. In your species this center is obviously the brain." The visitor tapped his head with his middle finger.

"Obviously," I readily agreed, tapping my head in fellowship.

"The fifth force is unrestricted by space or time: it acts instantly, by completely cancelling the distance between you, the emitter, and whatever point elsewhere in the cosmos towards which you have directed it. For instance, by activating the fifth force, you are able to see another world as clearly as if you were actually in it."

"I see." The most important thing in such conversations is to give the impression that you accept what you are being told easily and without skepticism. The more outlandish the matter, the more easily you should appear to go along with it.

"That is the idea that flashes. If you don't really know what's going on, that the fifth force has been activated, it will seem that you have made it all up, that nothing is real. But actually, nothing has been invented. The world that suddenly appears in your consciousness is no less real than your own, regardless of how unusual it may appear."

"Very interesting," I commented.

"All these books here are considered fanciful prose, while in my world they would be regarded as commonplace documents of unimpeachable authenticity. Your misconception will be rectified once you have mastered the fifth force, instead of using it in the wild, uncontrolled manner you have until now."

"If I've understood properly, then this would no longer be a bookshop but some sort of...archive?"

"Yes, a place where data about other worlds are collected, stored and made available. That is my field of work. I use the fifth force to investigate other worlds and catalogue them. That is how I came across the Earth."

"And so you decided to visit us?"

He shook his head abruptly. "No, no, you don't understand. It wasn't that simple. The fifth force does not transport matter to distant places. Only information. Whoever uses it does not move from his own world."

"But you've come here to Earth, right?"

"That happened because of the interference."

"Interference?"

"Yes. When two fifth force beams overlap."

"Aha, so that's it."

The visitor did not continue right away. He took out his handkerchief again and wiped his face. Several streaks of sweat were now streaming down his forehead, winding their way downwards to lose themselves in his beard. The vegetable smell emanating from him had become more powerful in the course of our conversation, almost intoxicating.

"When I directed my beam towards Earth, something highly unexpected happened. Another beam was heading outwards from here in the opposite direction at the same time. Someone had just flashed an idea about my world. A writer of science fiction, obviously, using the fifth force quite unskillfully, because if he knew the slightest thing about it he would never have let it happen. He would have known how dangerous it is when two beams interfere with each other."

"Dangerous?" I replied, properly aghast.

"Quite so. Two beams that interfere create a gap in the space-time continuum. If this gap is not quickly closed, it will start to suck in everything around it. First of all its two end points, Earth and my world in this case, then the planetary systems to which they belong, and then neighboring star systems. There is actually no end to its voracity. It's as though a black hole has opened up, eleven and a half light years long!"

I could only express appropriate horror. "Why, that's terrible! Horrible! Is there anything that can save us, or are we doomed to annihilation?"

"Yes, there is, if I am able to cancel the interference. It's still not too late for that. But time is running out."

"Then you must not hesitate," I said in haste. "How do you cancel the interference? What needs to be done?"

"I have to find the work about my world. Then go back with it and join it to my documentation about Earth. When these two fifth force products are joined together, the interference will disappear and the gap will close."

"But how will you go back? Please don't reproach me, but I still don't understand how you got here." This was not exactly in the spirit of my strategy. I usually avoid unnecessary questions, if for no other reason than because they are quite likely to be answered, which needlessly prolongs the conversation. But I felt I owed it to this eccentric somehow. He had taken pains to invent an admirable story, not some tedious inanity like most of the others. Many science fiction writers would envy him for this.

"Through the gap, of course. It can be used as a shortcut until it slips out of control. The crossing is instantaneous. I traversed all those light years in just one move, ending up in front of your bookshop. It was like stepping through to the other side of a kind of mirror, which was a new and very unusual experience even for me. I never thought I would ever go through a fifth-force interference zone. It may not look that way to you, but I am really no adventurer. Although I spend most of my time investigating other worlds, this is the first time I have physically left my own. Actually, I think I am more of what you would call a bookworm."

A rather uncomfortable smile appeared on the man's lips, as though in apology. I returned his smile, feeling suddenly sympathetic towards him. In other circumstances, this could have been an interesting exchange of ideas between two fellow writers, even somewhat kindred souls. I really liked his story. Even the bit about the shortcut wasn't bad. Not exactly original, but convincing nonetheless. As far as I could see, there was only one weak spot in the whole thing. I could have ignored it, but the hairsplitting critic in me prevailed in the end.

"I had no idea," I said, "that there were humans on other worlds, too. Yet so you must be—at least, to judge by your appearance."

"Of course there aren't."

"Well, then, how...?" I asked, indicating his body with my hand.

"Transformation," he replied succinctly, as though this explained everything.

"Ah, of course. I should have thought of it. Under the influence of the fifth force, indubitably."

"That's right. It makes it possible, while it is in interference, if you know how to manage it properly. But only for a short period. That is another reason why I am in a hurry. I won't be able to stay in this shape much longer. And I don't feel very happy in it. It's very uncomfortable and clumsy. I don't envy you this body one bit. It's extremely unsuited for movement, in particular."

"Surely there must have been some reason why you couldn't appear here in your own body?"

"Of course. I would die within moments. This is an extremely poisonous atmosphere for me, and the pressure is very high. Rarely have I come across such a dangerous environment, and I am acquainted with a very large number of worlds. But even if the conditions on Earth were perfect, I would still have to take human form. Because of you."

"Because of me?"

A smile played on the visitor's lips again. "Yes, because of you. How do you think you would have reacted if I had appeared in your bookshop in my natural form? Would you be conversing so casually with a ball?"

"A ball?" I repeated. A bell rang softly somewhere in the back of my mind.

"Yes, a ball, perfectly round and soft. What shape is more suitable than a ball in a world completely devoid of uneven spots and obstacles, and covered with dense vegetation? It's almost as if the entire planet were enveloped in a gigantic plant carpet. There is nothing lovelier than rolling on it."

I tried to swallow the lump in my throat, but my throat had suddenly tightened. I could feel my pulse start to pound dully in my ears.

"And what a captivating smell it has! That's what is actually the worst thing about Earth. I could somehow become accustomed to all the rest, but never this foul odor." He sniffed the heated air of the bookshop with disgust. "If you ever had the chance to smell the fragrances of my world, you would never be able to stand it here again."

I feverishly started to think. This wasn't really happening. It could not be happening! There must be some simple explanation. But none that crossed my mind made any sense.

"Smells," the visitor continued inexorably, "that emanate from the diversity of grasses that do not exist on any other of the multitude of worlds I have encountered to date. Lomus, rochum, mirrana, hoon, ameya, oolg, vorona…"

"…pigeya, gorola, olam," I continued with a voice deadened almost to a whisper.

The visitor's face lit up. "So that means you recognize the work!" he cried.

I recognized it, of course. It was truly a new story. So new that it had not yet been published, and thus could not possibly be found on the shelf over there with the recently published works. It was a story that no one but its author should or could know about at this moment. A story that resided, saved several times too many, in the virtual space of my computer.

I nodded briefly, wordlessly.

"Please give it to me. Quickly! If I don't hurry it might be too late."

As I slipped a diskette into the computer with automatic movements and pressed the keys to copy it, questions teemed furiously in my head. But I knew

I would not ask any of them. Not only because there was no time left for him to reply, but also because I was not really prepared for the answers.

The visitor took the diskette that I handed him, examined it carefully as though his eyes could see into its contents, then glanced at me and smiled again. He didn't say a word. I tried to smile, too, but it looked more like a grimace.

He turned around and headed hurriedly for the door. A moment later he was swallowed up by the thick wall of fog.

I stood there for a long time, motionless, staring at the impenetrable greyness that had engulfed him. And then my fingers hit the keyboard again. The tangle of letters disappeared from the screen in an instant, leaving behind a yellow void. The story that I had almost finished faded into nothingness. It left no trace behind it, just as the visitor had left no trace behind him. I could pretend to myself that I had never even written it, and that, as on so many other evenings, no one had entered the bookshop once the wispy spirits had made their sluggish ascent from the riverbed.

But I was deprived of this privilege to delude myself. The story had, in fact, been removed—just one erasure had destroyed all earlier saves—but the visitor had left a trace behind him after all. It was very faint, yet undeniable. I noticed it the first time I breathed in deeply through my nose. A tangle of delicate vegetable smells of unknown origin hovered faintly all around me. It might be impalpable to other people, but as long as I could smell it I knew I must restrain myself from writing science fiction.

7

The Puzzle

Mr. Adam only started to paint late in life, after his retirement. It happened quite unexpectedly. For the first sixty-five years of his life he had never shown any predisposition towards painting, for which he had neither talent nor interest. The arts in general attracted him very little.

The only exception might have been music, although he didn't really enjoy it. Sometimes he would find a radio station devoted mainly to music and leave it on low, just enough to dispel the silence that surrounded him during his long, dreary hours at work. It didn't matter what sort of music was being played; almost any would serve his purpose equally well, although he preferred instrumentals since singing distracted him. All he did at home was sleep, and often not even that, so there was little opportunity for anything else.

Retirement brought Mr. Adam an abundance of empty hours which he must fill. Experience gained at work had taught him that whenever he had to wait an indeterminate time for something, he had to impose obligations upon himself, and then discharge them doggedly, regardless of how unusual they might seem. This at least gave a semblance of meaning to everything. And one could not live without some meaning, however illusory.

He set himself one obligation for every day of the week. On Sunday he cooked, something he had never done before. He bought the biggest cookbook he could find in the bookstore and set himself to prepare every dish in it, in alphabetical order. The uncertainty of how far he dared hope to get at this

"The Puzzle." Written in 2001. Originally published in Serbian in 2000 as "Slagalica" in *Sedam dodira muzike/Seven Touches of Music*, Polaris, Belgrade, Serbia.

© Springer International Publishing AG, part of Springer Nature 2018
Z. Živković, *First Contact and Time Travel*, Science and Fiction,
https://doi.org/10.1007/978-3-319-90551-8_7

tempo did not disturb him. He was aware that he would require extreme longevity to reach the end of the book, but that was of no importance to him.

He followed the instructions for each recipe to the letter, and the only trouble he encountered was when they were not specific enough, but allowed the cook to use his own judgment or taste. He did not like everything he cooked, but that did not bother him greatly. He ate his culinary creations down to the last spoonful, throwing nothing away. This was almost a matter of honor to him. Sometimes, when the recipe was intended for several people, he ate the same food the whole week through.

On Monday Mr. Adam rode his bicycle. This was also a new departure. He learned how to ride easily and quite rapidly, despite his advanced age. He was not deterred by bad weather, though he would dress accordingly. The only trouble he had was when the rain spattered his glasses, unpleasantly fogging his vision. He preferred to ride without glasses in a downpour, though that rendered his vision equally foggy.

He always took the same route, each time increasing the distance a little. He tried to conserve his energy so he had enough left to go back by bike. He was only forced to return by other means of transport on the few occasions when there was a sudden turn in the weather, or he was overcome by fatigue. His conscience always plagued him when he gave up like that.

Unlike cooking, cycling had its limits. The route he took never actually ended, since it connected to many others, but even if he were to ride the whole day without stopping, which was not very likely, at midnight he would be required to stop. Tuesday was not for bike riding, but imposed its own obligation.

While still employed, he had read very little except professional journals. Not because there was no opportunity—many of his colleagues read for pleasure to pass the time at work—but because it seemed to him a sign of insufficient dedication to the job. Of course, his work would not have suffered for it, particularly since computers had taken over the bulk of his responsibilities. Now he decided to make up at least partially for this lapse. He became a member of the town library and went there every Tuesday. He entered as soon as it opened and stayed until it closed, only taking a short break early in the afternoon to eat something.

His initial subject was science fiction. This was a natural choice, but Mr. Adam soon gave it up. What he read about first contact seemed unsophisticated for the most part, often to the point of inanity—pulled out of thin air, at best. The number of writers demonstrating any knowledge of the real state of things was quite small, though such knowledge was easy enough to obtain. Disappointed, he was briefly tempted to abandon reading entirely. But giving up in the face of adversity was not in his nature, and besides, he had paid his dues a year in advance. Finally, were he to stop going to the library he would have to think up a new obligation for Tuesday, and that prospect did not please him at all.

He found a solution to this problem, using the same means he had often resorted to at work. Whenever his search in one area drew a blank, he simply broadened his field of vision. Not knowing what else to choose, this time he broadened the field to the farthest limit, like suddenly taking the whole sky instead of one small sector. Instead of science fiction he chose literature in its entirety, but as this turned out to be far greater even than the cookbook, he had no idea at first where to begin.

The main catalogue was indexed by author, and he briefly considered adhering to that order. But then he thought again, and concluded that this would not be a satisfactory approach. He spent some time at the library computer, classifying titles by publication date, and finally obtained a list of books from the oldest to the most recent. The scale of this list did not discourage him at all—he had become accustomed to such challenges long ago. He started to read steadily, without rushing, as if all the time in the world lay before him.

On Wednesdays Mr. Adam went to the zoo. The middle of the week was the right time to visit, when there were far fewer visitors than at weekends. Moreover, if the weather was bad, he would often see no one in the vicinity for long periods. That suited him best. Ideally he would have liked to be completely alone at the zoo, but of course, he was never able to count on that.

Mr. Adam did not behave like the ordinary sort of visitor, who just wanders around enjoying himself. First he found out which animals were housed in the zoo, then he drew up a schedule of visits. Each animal was allotted a whole day. Few of the zoo's inhabitants were worthy of such dedication, but the systematic patience with which Mr. Adam approached everything did not allow him to act otherwise.

He would arrive in the morning at the chosen cage and sit in front of it. When there was no bench he brought a small folding chair from home. He would stay in that spot until nightfall, doing nothing but observe the animal carefully through the bars. He did not know exactly what to expect. Certainly nothing special. What he hoped for was at least a certain reaction to his presence, just an awareness that he was there, perhaps a glance that deliberately crossed his own. Anything short of complete disregard.

It was actually quite easy to attract the animals' attention by offering them food, but Mr. Adam never did. It would be a form of cheating, and he would brook no cheating. Therefore he took no food with him, not even for himself. When he left the zoo on a Wednesday evening, he was often faint with hunger.

On Thursdays Mr. Adam visited churches. Not being religious, he had never been to such places before, and was surprised to learn that the town held sixteen of them. Sometimes he had to walk the whole day in order to take them all in. He could have used public transport, of course, which would have sped things up considerably, but that would have run contrary to Mr. Adam's basic intention. His Monday bike ride was by no means sufficient to keep him in shape, and his need for additional exercise was the more acute after spending all Wednesday sitting still at the zoo. What could be more appropriate than a seriously long walk?

In order to avoid the tedium of repeating the same walk every time, Mr. Adam took a different route every Thursday. This was not done at random; he had worked out a precise plan. He approached it as a simple problem in combinatorial mathematics. There were far more ways of ordering the sixteen points than he imagined he would ever need. The itineraries greatly varied in length, because the algorithm he had chosen took no account of the distance between the churches. He bore up stoically under this inconsiderate mathematical dictate, consoling himself with the reflection that he found longer walks more enjoyable.

Mr. Adam could have visited points other than churches. In principle, the direction of his walks was immaterial to him, so he could not have explained why he had made churches his choice. Luckily, no one ever asked him, which saved him from embarrassment. On reaching a church he began by walking all the way round it, examining it inquisitively, as if seeing it for the first time. Then he would take a little rest, sitting in the churchyard if there was one, before continuing on his way.

In time he got to know the exteriors of all sixteen churches quite well, and came to regard himself as a real expert in this field. He believed that he alone had noted some of the details. For example, there was always an even number of birds' nests under the eaves. Who knows why? He rarely felt any urge to examine the churches' interiors. He was only tempted to enter on two or three occasions, but he always refrained, and here again he was unable to say what it was that had dissuaded him.

Friday was his day to go to the movies. Mr. Adam would always watch four films in a row, from mid-afternoon to late in the evening. This was by any standard too much. After the second film his impressions were already becoming confused, and by the end of the fourth he would feel truly exhausted, as though he had been working at some strenuous task, rather than sitting in a comfortable seat the whole time. But this did not prompt him to decrease the number of films.

Mr. Adam was not the least bit selective regarding the repertoire. He did not have a favorite film genre, although he felt most relaxed watching romantic comedies. Action films left him rather indifferent, and although they were loud as a rule, he even managed to doze off to them, particularly if they were the last of that day's four. He found thrillers unconvincing, although not as much as most science fiction films. Those sometimes appeared outrageously idiotic; he could never understand why filmgoers got so excited about them. Overly erotic scenes embarrassed him, but fortunately that was not noticeable in the dark.

Although it might have appeared that Mr. Adam chose his films at random, this was not at all the case. He bought his tickets with great care, concentrating on films that were expected to sell out. Just before the lights dimmed, Mr. Adam would stand up for a moment and look all around. He would feel annoyed should he spot any empty seats. Those empty places would bother him until the end of the show. He only felt at ease in a full house. That alone could temporarily lighten the burden of solitude which, like some sinister inheritance, hung over from his former work.

Mr. Adam passed Saturday in the park. He needed to spend time outside in the fresh air after so many hours indoors the previous day. Late in the morning he would go to the large city park with its pond in the middle, and head for the bench where he always sat. On the rare occasions when someone was already sitting in the place he considered his own, on the far left-hand end of the bench next to the wrought iron armrest, Mr. Adam would wait unobtrusively to one side for the bench to come free. It did not bother him if the remainder of the bench was occupied, though he avoided entering into conversation with strangers.

On warm, sunny days he would stay there until dusk, doing nothing but idly watching what was happening around him: people strolling by, dogs chasing each other frantically on the grass, leaves rustling in the surrounding treetops, birds gliding silently through the blue sky, sudden ripples on the smooth surface of the pond. Until recently this idleness would have seemed an extremely foolish waste of time. Now, however, the tables were turned. He saw everything which had gone before as a waste of time. All his previous life. All the years, all the effort, all the hopes.

That was not how it had seemed, at any rate not in the beginning. Not in the least. It was a pioneering time of great excitement. Great expectations. And great naiveté. They thought that contact was only a matter of time. The cosmos was teeming with life, messages were streaming between worlds, all that was needed was to prick up our electronic ears to hear them. Without this optimistic certainty the money for the first projects would never have been found—investments that could pay off stupendously as soon as the inexhaustible wealth of knowledge started to pour in from the stars.

Mr. Adam had fond memories of those early days, despite later disappointments. There was something romantic in the anticipation that overcame him whenever he put on his earphones. He spent countless hours listening to the cacophony streaming from the skies, straining to recognize some sort of orderly system in it. Like all his colleagues, he secretly hoped that he would be the first to hear the signal.

But as time passed and nothing arrived except inarticulate noise, the true proportions of the task started to emerge. Since listening to the closest star systems produced no results, there was a shift to more distant ones, but each new step brought a substantial increase in their number. The initial enthusiasm foundered when it was established that more than one generation might be needed to complete the task. This led many people to leave the search for extraterrestrial life in favor of more promising areas, and financiers were less and less willing to continue investing in something so vague and unreliable.

Fortunately, at that point computers were introduced, with their numerous advantages over people: they are incomparably faster, more effective and dependable, and do not quickly lose heart in the face of failure. Even so, Mr. Adam did not look upon their use with total approval. Computers reduced people to commonplace assistants whose sole purpose was to serve them. What had begun as a noble project for the chosen few degenerated into a routine technical duty that almost anyone could perform—mere waiting, leached of any true excitement. The last remnants of romance vanished without a trace.

After several decades had passed, and the computers had meticulously checked many millions of sun systems but detected no sign of extraterrestrial intelligence, Mr. Adam felt a certain gloomy exultation. His feelings were paradoxical, because only under opposite circumstances, with contact made, would he be able to say that his life's work had meaning. On the other hand, contact achieved with the assistance of computers would to him be some sort of injustice, almost an anticlimax.

Despite the silence of the cosmos, the search programs were not discontinued. Although large, the number of investigated stars was trifling compared to the total number of suns in the galaxy. In principle, one of the giant radio telescopes could start receiving the long-awaited message from the very next spot in the sky. However, as his retirement approached, Mr. Adam became more and more skeptical in this regard.

It was not just the realization that the prospects of finding Others within his lifetime were negligible; he could somehow reconcile himself to that if he was sure they were on the right track. But the suspicion started to trouble him that the reason for failure lay not in the fact that only a tiny part of the sky had been investigated, rather in something much more fundamental. What if some of the basic assumptions upon which the entire project was founded were wrong?

Maybe there was no one out there after all. Maybe sentient beings were so unlikely that they had only appeared in one place. Everyone was convinced of the opposite, but this conviction had no solid basis. Behind it might lie an unwillingness to accept the terrifying fact of cosmic solitude. As the years passed, Mr. Adam started to feel anxious under the unbounded wasteland. The starry sky pressed heavily upon him at times. The strange need arose for some sort of shelter, for consolation.

Suppose extraterrestrials exist and are communicating, but we don't recognize it? What if they were doing it in some other way, and not the way we presumed? Mr. Adam had never asked himself this question seriously. Whenever it stole quietly into his consciousness he would expel it hurriedly, with a sense of hostility and guilt, as any true believer rejects a heretical thought. All his sober, scientific being opposed it. Similar inconsistencies had prevented him from coming to like science fiction.

He still considered this the proper approach, despite all the unfulfilled hopes in the life that yawned behind him. And at the end of the day, what other means besides electromagnetic waves could be used to communicate between the stars? With regard to his past, the daily obligations he set himself helped put it out of his mind. Perhaps these obligations really were meaningless, but the problem of meaning no longer plagued him. He enjoyed everything he was doing now, even idling in the park each Saturday, and that pleasure was all that mattered. In any case, he was not just idly passing the time. He had recently started to paint.

Music had been the catalyst. Upon reaching the park one Saturday at the beginning of summer, he found that a bandstand had been erected near his bench. It had not been there seven days previously, nor had anything heralded its advent. This had irritated Mr. Adam no end. Although pretty, with its slender columns and domed roof, he considered it an unconscionable desecration of the environment. In addition, the bandstand largely blocked his view of the pond, and he seriously considered looking for another place to sit. But habit won out and he stayed on his bench, scornfully endeavoring to disregard the interloper.

This ceased to be possible when musicians climbed onto the bandstand at noon. They were formally dressed and the conductor even wore a tuxedo with a large white flower in his lapel. They sat on chairs placed in a circle and spent some time tuning their instruments. Mr. Adam found this dissonance an additional nuisance. It not only sounded awful but started to attract park visitors, and rather a large crowd soon formed. A crowd of people, however, was the last thing Mr. Adam wanted after his Friday spent in a packed movie theatre.

He would have to move after all. He couldn't stand this. But just as he started to rise the music began. He stopped halfway, transfixed, then slowly sat down again on the bench. All at once he was no longer surrounded by too many people, his bad mood disappeared, and nothing existed beyond the music. He stared fixedly at the bandstand, immobile, listening intently.

This paralysis did not last long. He came out of it suddenly and began feverishly rummaging through his jacket pockets. It seemed to take forever to find what he was after. He always carried a notebook and pen with him. Since retirement he had not written anything in it, but he carried it with him nonetheless. He opened it hurriedly and started to draw. He dared not miss a thing.

He drew short, brusque lines, just like a stenographer taking rapid dictation. The pages in the notebook were small, so he filled them quickly. He was afraid he would run out of pages before the music ended, but fortunately the notebook was thick enough. Even so, he made the last drawing on the brown cardboard covers. Had the music lasted a moment longer, there would not have been enough room. The very thought suddenly filled him with horror.

The listeners' echoing applause after the last chords had the effect of an alarm clock suddenly going off. Mr. Adam jerked like one waking from restless sleep; he turned this way and that in confusion for several moments as if trying to figure out where he was. He feared he would arouse the suspicion of those around him, but no one paid any attention to the old man on the end of the bench, engrossed in his writing. All eyes were turned toward the conductor who was bowing theatrically.

Mr. Adam stood up and walked away unobtrusively. There was no reason to stay there any longer. During his extensive walks between churches he had come to know the town quite well, so he knew exactly where to find a shop with painting supplies. There might have been one closer, but he would waste more time inquiring after and finding it than it took to reach the other. The salesman noted with a smile that he was clearly preparing a serious project, judging by the quantity of materials he had purchased. Mr. Adam returned the smile, mumbled something vague, then hurried home.

Unskilled at painting, he had trouble setting up the easel properly, but then got down to work. He opened the notebook and began carefully transferring onto the canvas what he had written, as if neatly copying over rough notes taken in a hurry. He worked slowly but with passion, unaware of the passage of time. When he had finished it was already quite dark.

He did not know what he had painted. Viewed from up close it looked just like random strokes of paint. He was convinced, however, that not a single stroke of the brush had been accidental, that everything was exactly as the music ordered, in spite of his inexperience. When he moved back from the painting a bit, he thought he could make out part of a larger shape, but he wasn't sure. It suddenly crossed his mind that before him was just one piece of some larger puzzle. He thought briefly about what to do with the canvas, and then he hung it unframed on one of the bare walls.

The next Saturday he went to the park well prepared. He no longer needed the notebook as intermediary. He sat at his usual place on the bench and set up the easel in front of him, holding paintbrush and palette. In different circumstances he would have abhorred the inquisitive peering of bystanders, although a painter at work was certainly not unusual in the park. Now, however, he paid no attention, concentrating exclusively on the impending concert.

This time he painted rapidly. It lasted just as long as the music. When the applause resounded, Mr. Adam, panting and sweaty, had just finished covering the last white space with paint. Before the crowd dispersed, several pairs of eyes glanced at the painting, perplexed, since it did not depict anything recognizable. A short, elderly woman dressed in a bright orange dress stopped by the bench for a moment. She took an enormous pair of glasses out of her handbag and examined first the painting and then the painter. "Very nice," she said with a smile. She put her glasses back in her handbag, nodded in brief approval and walked away.

As a man unaccustomed to compliments, Mr. Adam felt ill at ease. The woman's words were by no means unpleasant, quite the contrary, yet he was still glad she had not lingered. He would have been in the awkward situation of having to say something in reply. He waited a while for the elderly woman to move on, then collected his equipment and hurried home. He could have stayed in the park longer, his work was completed and the day was very fine, but curiosity got the better of him.

He put the new canvas next to the other one on the wall. He had no expectations and thus was not very disappointed when it turned out they had no points in common. For a moment, though, he thought he could make out some part of a greater whole in the second painting, too, but here again it was most likely just his imagination. In the absence of any recognizable form he thought he saw something that was not actually there. This was a trap he had learned to avoid back in the early period, before computers, while listening to the stars with his own ears. If you're expecting a horseman you have to be very careful not to mistake your heartbeat for the beat of a horse's hoofs.

The next fourteen Saturdays, all summer long, each time Mr. Adam returned from the park he had one more painting to place on the wall next to the others. In time his brisk, almost frenetic painting became something of an attraction at the park, and a good many music-lovers would stand around to watch him work. He paid no attention to them. At the end of the music and painting he would quickly glance through those gathered around him, but never once did he catch sight of the slight figure in orange.

When Mr. Adam reached the park on the first Saturday in September, carrying his painting materials as usual, a surprise awaited him. The bandstand had disappeared as unexpectedly as it had arrived. It had been removed very carefully, leaving no trace behind—not even trampled grass. He darted in bewilderment around the spot where the little structure had stood, overcome by completely opposite feelings from those which had assailed him in the beginning. Now he missed the bandstand, and the environment seemed somehow naked and incomplete without it. For a moment he considered inquiring as to why it was no longer there, maybe even lodging a complaint, but he did not know where this should be done and in the end dropped the idea.

He returned home in a dejected mood and sat in the armchair facing the wall covered with paintings. The canvases formed a large square: four paintings in each of four rows. He stayed there for seven full days, only leaving the armchair to take a quick bite or go to the bathroom. He even slept there in his clothes, but the brief, restless, erratic sleep did not refresh him. He changed the distribution of the paintings from time to time. During that long week filled with almost constant pouring rain, he tried just a tiny fraction of all possible combinations of the sixteen canvases.

On the evening of the following Saturday he got up from the armchair, stretched, and went to the window. Rays from the low sun in the western sky were cutting a path through patchy clouds, just like gleaming swords. He stayed there a while looking absently at the flickering play of light. Then he went to the wall and took down the paintings. He couldn't carry them all at once and had to make two trips to the basement, where he left them.

When he came up from the basement the second time, he went into the kitchen, took the large cookbook down from the shelf, opened it at the bookmark and became immersed in reading the recipe that was next in line. The following day was Sunday, his cooking day.

8

Time Gifts

The Astronomer

I

He had to escape from the monastery.

He should not have been there at all; he had never wanted to become a monk. He'd said so to his father, but his father had been unrelenting, as usual, and his mother did not have the audacity to oppose him, even though she knew that her son's inclinations and talents lay elsewhere. The monks had treated him badly from the beginning. They had abused and humiliated him, forced him to do the dirtiest jobs, and when their nocturnal visits commenced he could stand it no longer.

He set off in flight, and a whole throng of pudgy, unruly brothers started after him, screaming hideously, torches and mantles raised, certain he could not get away. His legs became heavier and heavier as he attempted to reach the monastery gate, but it seemed to be deliberately withdrawing, becoming more distant at every step.

And then, when they had just about reached him, the monks suddenly stopped in their tracks. Their obscene shouts all at once turned into frightened screams of distress. They began to cross themselves feverishly, pointing to something in front of him, but all he could see there was the wide open gate and the clear night sky stretching beyond it. The gate no longer retreated before him, and once again he felt light and fast.

"*Time Gifts.*" Written in 1997. Originally published in Serbian in 1997 as *Vremenski darovi*, Polaris, Belgrade, Serbia.

He was filled with tremendous relief when he reached the arched vault of the great gate. He knew they could no longer reach him, that he had gotten away. He stepped outside to meet the stars, but his foot did not alight on solid ground as it should have done. It landed on something soft and spongy, and he started to sink as though he'd stepped in quicksand. He flailed his arms but could find no support.

He realized what he had fallen into by the terrible stench. It was the deep pit at the bottom of the monastery walls; the cooks threw the unusable entrails of slaughtered animals into it every day through a small, decayed wooden door. The cruel priests often threatened the terrified boy that he, too, would end up there if he did not satisfy their aberrant desires. The pit certainly should not have been located at the entrance to the holy edifice, but this utmost sacrilege for some reason seemed neither strange nor unfitting.

He began to sink rapidly into the thick tangle of bloated intestines, and when they almost reached his shoulders he became terror-stricken. Just a few more moments and he would founder completely in this slimy morass. Unable to do anything else, he raised his desperate eyes, and there, illuminated by the reflection of the distant torches, he saw the silhouette of a naked, bony creature squatting on the edge of the pit, looking at him maliciously and snickering.

He did not discern the horns and tail, but even without these features he had no trouble understanding who it was; now that it was too late, he realized whom the terrified monks had seen. He froze instinctively at this pernicious stare, wishing suddenly to disappear as soon as possible under the slimy surface and hide there. All at once the blood and stench no longer made him nauseous; now they seemed precious, like the last refuge before the most terrible of all fates.

And truly, when he had plunged completely into that watery substance, it turned out that it was not, after all, the discarded entrails of pigs, sheep, and goats, as it had seemed to be, but was a mother's womb, comfortable and warm. He curled up in it, knees under his chin, as endless bliss filled his being. No one could touch him here; he was safe, protected.

The illusion of paradise was not allowed to last very long, however. Demonic eyes, like a sharp awl, quickly pierced through the layers of extraneous flesh and reached his tiny crouched being. He tried to withdraw before them, to retreat deeper into the womb, to the very bottom, but his persecutor did not give up. The thin membrane that surrounded his refuge burst the moment he leaned his back against it, having nowhere else to go, and he fell out—into reality.

And with him, out of his dream, came the eyes that persisted in their piercing stare.

He could not see them in the almost total darkness, but their immaterial touch was nearly palpable. Suddenly awake, he realized that someone else was with him in the cell. He had not heard him come in, even though the door squeaked terribly, since probably no one had thought to oil it in years. How strange for him to fall into such a deep sleep; the night before their execution, only the toughest criminals managed to do that. They were not burdened by their conscience or the thought of impending death, and he certainly was not one of them.

He raised his head a bit and looked around, confused. Although he felt he was not alone, his heart started racing when he really did see the shape of a large man sitting on the bare boards of the empty bed across from him. If not for the light from the weakly burning torch in the hall that slanted into the cell through a narrow slit in the iron-plated door, he would not have been able to see him at all. As it was, all he could make out clearly were the pale hands folded in his lap, while his head was completely in shadow, as though missing.

He asked himself in wonder who it could be. A priest, most likely. They were the only ones allowed to visit prisoners before they were taken to be executed. Had the hour struck already? He quickly looked up at the high window with its rusty bars, but there was no sign of daybreak. The night was pitch black, moonless, so that the opening appeared only as a slightly paler rectangle of darkness against the interior of the cell.

He knew they would not take him to the stake before dawn, so he stared at the immobile figure uncertainly. Why had he come already? Would they be burning him earlier, perhaps, before the rabble gathered? But that made no sense. It was for this senseless multitude that they organized the public execution of heretics, to show in the most impressive manner what awaited those who dared come into conflict with the catechism. The sight of the condemned, his body tied or nailed to the stake, writhing in terrible agony while around him darted fiery tongues of flame, had a truly discouraging effect on even the boldest and most rebellious souls.

Or maybe this was a final effort to get him to renounce his discovery. That would be the best outcome for the Church, of course, but he did not have the slightest intention of helping it; on the contrary, had he come this far just to give up now? If that was what was going on, their efforts were in vain.

"You had a bad dream," said the unseen head.

The voice was unfamiliar. It was not someone he had already met during the investigation and trial. It sounded gentle, but this might easily be a trick. He was well acquainted with the hypocrisy of priests. His worst problems had been with those who seemed understanding and helpful and then suddenly showed their pitiless faces.

"Why do you think that?" asked the prisoner, stretching numbly on the dirty, worn blanket that was his only bedding.

"I watched you twitch restlessly in your sleep."

"You watched me in the total darkness?"

"Eyes get accustomed to the dark if they are in it long enough, and can see quite well there."

"There are eyes and eyes. Some get accustomed to it, others don't. I ended up here because I refused to get accustomed to the dark."

The fingers in the lap slowly interlaced, and the prisoner suddenly realized that they looked ghostly pale because he was wearing white gloves. They were part of the church dignitaries' vestments, which meant that the man in the cell with him was not an ordinary priest who had been sent to escort him to the stake. So, it was not time yet.

"Do you think that you will dispel the darkness with the brilliance of your fiery stake?" The tone was not cynical; it sounded more compassionate.

"I don't know. I couldn't think of any other way."

"It is also the most painful way. You have had the opportunity to witness death by burning at the stake, isn't that right?"

"Yes, of course. While I was at the monastery they took us several times to watch the execution of poor women accused of being witches. It is a compulsory part of the training of young monks, as you know. There is nothing like fear to inspire blind loyalty to the faith."

"Yes, fear is a powerful tool in the work of the Church. But you, it seems, have remained unaffected by its influence?"

The prisoner rubbed his stiff neck. He could still somehow put up with the swill they fed him, the stale air and the humidity that surrounded him, and the constant squealing and scratching of hungry rodents that he'd been told were liable to bite the ears and noses of heedless prisoners. But nothing had been so hard in this moldy prison as the fact that he did not have a pillow.

"What do you expect me to answer? That I'm not afraid of being burned? That I'm indifferent to the pain I'll soon be feeling at the stake? Only an imbecile would not be afraid."

"But you are not an imbecile. So why didn't you prevent such an end?"

"I had no choice."

"Of course you did. The only thing you were asked was publicly to renounce your conviction and to repent, which is the most reasonable request of the Court of the Inquisition when serious heretical sins are involved. If you had done that, you would have kept your title of royal astronomer and been allowed to continue teaching students."

"Who would attend the lectures of a royal astronomer who had renounced his discovery out of fear?"

"There is a question that comes before that. Why did you have to announce it in the first place? What did you want to achieve by that?"

"What should I have done—kept it a secret, all for myself?"

"You were aware that it goes counter to the teachings of the Church. You should have expected her to take measures to protect herself."

"Of course I expected that. But I was relying on her hands being rather tied."

"It doesn't look that way, judging by the sentence you were given."

"Oh, you know perfectly well that the stake is not what the Church wanted. It was a forced move after all attempts to talk me into cooperating failed."

"Based on your condition, I would not say that they tried all possible means. You do not look like someone who has been given the Inquisition's full treatment."

"Well, I'm not a witch. They didn't have to force me to agree to some meaningless accusation. I did not deny my guilt. That is why the whole investigation proceeded like some kind of friendly persuasion, even though, probably just to impress me, in the background stood the power of all the devices to mutilate, quarter, cut, break, and crush. But I was not even threatened with one of them, let alone put to any device. You do not torture someone who is valuable to you only as an ally. What good would it be if the royal astronomer were lame or blind?"

"Not even after the alliance has been irrevocably called off? The Inquisition can hardly boast of the virtues of forgiveness and compassion."

"That is why it is renowned for its patience and acumen. The sentence was passed, but I have not been burned yet. There is still time. Attempts to win me over to the Church's side will continue to the very end. In any case, that is why you are here, isn't it?"

There was an indistinct commotion from the end of the hall, followed by the sharp sound of a key unlocking a door and someone groaning painfully as he was thrown into a cell like a bag of potatoes. The Inquisition's investigators did their work primarily at night. The main interrogation room was in the basement; in spite of the thick walls, horrible screams could be heard periodically, weakening the last remains of will and resistance in the other prisoners awaiting their turn to be taken down there. As they moved off after closing the door with a bang, one of the guards muttered something to the other, making him laugh raucously. For a long time his burst of laughter echoed like thunder through the stone hallway.

"But you, of course, will not relent?" asked the voice from the darkness after the echo finally died out.

"Of course."

"What is the real reason for that?"

"What do you mean?"

"You certainly are not a simpleminded idealist who has gotten involved in all this because you don't understand how the world works, what forces set it in motion. On the contrary, everything you have done from the beginning seems to have been carefully planned. You have lit a fire that only you can put out. It takes great resourcefulness to turn the tables on such an experienced service as the Inquisition, to tie its hands, as you say. And it takes the courage of a fanatic that is always lacking in idealists at the crucial moment, the readiness to go all the way, no matter what the cost. You, naturally, shy away from the pain that awaits you at the stake, but you will go to your execution nonetheless just because that will harm the Church the most. What is it that she has done to you?"

The prisoner started to rise into a sitting position on the hard bed, feeling a stab of pain run all the way down his stiff back. As he did so, a scene from his dream suddenly rose to the surface of his memory. It was very vivid, although fixed, like some sort of ugly picture: the twisted faces of the monks lustfully reaching for his tiny, helpless body.

"Isn't it still early for my last confession?"

"I'm not here to listen to your confession."

"Oh, yes, I almost forgot. You are here to prevail upon me to change my mind. But if you truly believe what you just said, it must be clear to you that it's impossible."

"It is clear to me."

"Then why are you wasting your time?"

There was no immediate reply from the other side of the cell. A hand rose and reached for something that was lying unseen on the wooden bench. A moment later it returned to the flickering shaft of light from the torch in the hall. It was now holding a slender black cane with a carved white figure on the top.

"I have more than enough time." The voice seemed to become muffled, more distant.

"But I don't. My hours are numbered."

"That's right. Soon they will come to take you to the stake, but before that you will be given one last chance to accept the Church's offer. But, as we know, you will refuse. Although it makes no difference, really."

"It does make a difference. If I accept, everything I did will have been in vain."

"No, it won't. The damage was done the moment you announced your discovery, and it cannot be undone. The fluttering of the butterfly's wings should have been prevented before it initiated the storm. Even if the Church made a sincere ally out of you, it would only slow down the harmful repercussions."

"Do you really think that this is sufficient to make me change my mind? I expected you to come up with something more convincing."

"I have no intention of dissuading you. But that is the way things stand nonetheless. Heresy has been sown on fertile ground. Neither the stake nor repentance will turn your students away. They will start to spread forbidden knowledge, to add to it. Once set in motion, this course cannot be stopped, even though the Inquisition will take every measure to obstruct it. You have let the genie out of the bottle, and he can no longer return to it. The Church will finally realize this inexorability, but it will be too late then."

The prisoner strained to make out the hidden face in the impenetrable obscurity, but without success, even though his pupils were completely dilated.

"Isn't it unbecoming for a man of God to have so little faith in the future of the Church?"

"Why do you think I am a man of God?"

A shroud of silence suddenly descended on the cell. Several long moments passed before the prisoner realized what was wrong. He had spent many nights alone in this place, and he could always hear some sort of noise: moaning from one of the neighboring cells, the screech of rusty hinges, the murmur of the guards, muffled cries from the basement, the rustling of mice and rats, the creaking boards on which he lay, distant sounds of the outside world. Now all of that had mysteriously disappeared.

"Who are you?" he said, finally mustering the courage to break this tomblike silence. The darkness did not answer; suddenly, once again the prisoner felt the stab of the piercing eyes that had followed him out of his dream. "The Tempter?" The word was almost inaudible, so that he didn't know whether he said it or only thought it.

"Why should that bother you?" The voice remained just as gentle. "If I am the Tempter, then we are on the same side. We have the same opponent."

"Why...why are you here? What do you want from me?" He had a strong urge to cross himself but at the last moment thought it somehow inappropriate.

"I don't want anything from you. On the contrary, I have a gift for you. Sort of a token of our alliance. A trip."

"A trip?"

"Don't worry, you won't leave this cell, and you will get back on time, before they come for you."

"What kind of a trip will it be if I stay here?"

"The only one possible under the circumstances: through time."

The prisoner blinked. This was not really happening. He was still asleep. However, there was none of the awakening that necessarily followed such a realization. He brought his hand to his face and pinched his cheek hard. The pain was real. Only too real.

"I don't want...to go...anywhere."

"But you'll like it there. I'm quite sure. The future has pleasant surprises for you."

"The future?"

"Yes. Almost three hundred years from now."

"Why would I want to go...to the future?"

"Out of curiosity, above all. Aren't you interested in checking whether you really succeeded in outwitting the Church? Even though you certainly appear self-confident, there must still be a shadow of doubt in there. What if your sacrifice is in vain?"

"But you said it isn't. That my students..."

"A moment ago that did not sound convincing to you. In any case, can you believe in the word of the Tempter, even when you're on the same side as he is?"

"What would the future corroborate? What would I see there?" As he asked these questions, he felt completely foolish. He had let himself be drawn too easily into a crazy, impossible conversation. Where was the common sense he took such pride in? Had he gone out of his mind? He had heard that this sometimes happened to people waiting to be burned at the stake. Fear twisted their minds.

"A better question would be what you won't see. First of all, you won't see a monastery on the top of this hill. Its walls will still be there, but it will no longer contain dark, humid cells, corridors all sooty from torches, or a torture chamber in the basement."

"The monastery will fall into ruin?"

"No, it will be remodeled."

"What can you remodel a monastery into?"

The answer was preceded by a brief silence that seemed to indicate a certain hesitation, indecision. "I suppose that in the end you would recognize it without my help, although it will certainly look...strange. But I would do well to prepare you. You will not have much time, and the future can have a stunning effect. At the time of your visit, instead of a monastery this will be an astronomical observatory."

He knew that he should say something in return, that it was expected of him, but he could not utter a word. His vocal cords were vibrating, forming confused questions, but his throat had closed completely and no sound came out. He stared blankly ahead, his mouth a void.

In the infinite silence that reigned once more, a white-gloved hand set the cane between the knees, then disappeared in the folds of the black robe. The hand took a moment to find something there, then emerged with a round, flat object on its open palm. Golden reflections shone from its engraved curves. The dark figure's thumb moved along the edge of the object and the lid popped open.

The hand extended toward the prisoner, but he remained stock-still. It was not indecision; the spasm that had closed his throat had now spread to his entire body. He wanted to move, do something, anything, he couldn't stay there motionless forever, but his muscles refused completely to obey.

"Yes, before you leave, there is one more thing you should know. It will please you, I believe. The observatory will be named after you."

The movement with which he accepted the watch had nothing to do with his will. It seemed to him that someone else received the Tempter's gift, that he was just an observer who should in fact warn the incautious sinner not to do it, that it was insane. He wouldn't have listened, anyway, his soul was already lost, so it made no difference; actually, nothing could help him anymore.

The watch face radiated a bright whiteness. In the dark cell it was a lighthouse summoning sailors, the flame of a candle attracting buzzing insects, a star luring the glass eye of the telescope. And over it were two ornate hands at a right angle, forming a large letter L.

II

Staring at the shiny surface, he failed to notice the changes that had started to take place. Something sparkled in the cell, apparitions passed through it more transparent than ghosts, and the specter on the other bed instantly dissolved into nothingness.

His attention was attracted only by sudden daylight in the high barred window.

Isn't it still early? he asked himself, raising his eyes in bewilderment.

But the time of miracles had just begun. His eyelids barely had time to blink before it was dark in the window again. The astronomer in him opened his mouth to contend the obvious, but he was silenced by the stronger voice of the child who cares not at all whether something is possible or not, as long as it is fascinating.

Many short interchanges of light and darkness took place before the child had had enough of this monotonous kaleidoscope, finally letting the scientist think about solving the mystery. There was only one explanation, of course. To accept it, however, one had to accept the impossible almost as an act of faith.

Before him the days and nights were passing at accelerated speed, but he could not ask the questions dictated by his reason. He had lost that right the moment he took the watch. In any event, was the "how" important? If this was the way to travel to the future, so be it.

Finally the hypnotic flashes of blue-gray and black images in the stone window tired even the astronomer. He turned around—and at first it seemed that the dizzy rush through time had stopped. Nothing was moving, everything looked fixed, unchanging. And then he realized that it was only an illusion. There could be no rapid changes here: the monastery walls were built to withstand the centuries.

Nonetheless, there were a few things in the cell made of less durable material. He stood transfixed as he watched the boards on the bed across from him gradually swell up from the perpetual humidity and then split and fall to the ground, where they slowly turned into a shapeless mass on the flagstones.

He jumped up from his bed when it struck him that the same fate had to affect the boards on which he was sitting. Sure enough, they also ended up as a pile of sawdust. He, however, had not felt a thing: if this possibility had not crossed his mind, he would have continued to sit calmly on nothing, in midair.

The wooden door was considerably thicker, but in the end it, too, succumbed to the effects of decay. First the steel bars fell off, then the hinges gave way, cracks appeared, then gaps and holes, until finally there was nothing to stop him from going into the corridor. The cell ceased to be a prison. But on the other side of the threshold, freedom was an impenetrable darkness, since no one lit torches to dispel it anymore.

Thoughts of freedom reminded him of the many prisoners who must have sojourned here in misery after him. During this rapid movement through time he could not see them, of course, although here and there he had the deceptive feeling that there was someone else with him. During the instants of darkness that were nights, a shape seemed to bulge on the bed across from him, but this illusion was too brief to make anything of it. In the flashes of lightning that were days, something would flicker in front of him periodically, a certain hint of movement, but it was as cryptic as a flash seen out of the corner of the eye.

The ceiling disappeared so suddenly that he did not have time to catch his breath. It was there one moment and then suddenly gone without a trace, as though a giant had taken a huge lid off the monastery. At the same time, all the partition walls were removed, leaving only the solid outer walls that no longer had any windows.

The rapidly changing days and nights were incomparably more exciting with the entire firmament spread over his head than before, when he had only had a tiny corner of sky. The entire universe seemed to be hurriedly whispering some secret message to him...

But he was not given the time to figure it out. Just as mysteriously as the lid was lost, it returned a few moments later, although not the old one. He found himself inside an enormous closed space over which there rose a gigantic dome. Only cathedrals boast such roofs, he thought, but this was certainly not a cathedral; their domes did not have a wide slit cut through the center, let alone a large cylinder pointing upward through that opening.

He did not realize that the voyage was over because there was no slowing down; it happened all at once. He was looking at the empty opening in the vault over his head, but many heartbeats had to pass before he finally noticed that the alternating light and darkness had stopped. The night sky that settled in his eyes was sprinkled with the clusters of stars found in the thin air of mountain peaks.

A click in his hand jolted him out of the paralysis that had overcome him. The watch had completely slipped his mind, although it had been in his outstretched palm the entire time. Now it had closed, since its magic work was finished. He originally thought to put it in his pocket but then decided he should keep it in his hand; his first idea would have shown inadmissible disrespect.

He slowly and timidly began to turn around in the semidarkness of the large area. As wondrous things whose purpose he could not divine entered his field of vision, he remembered the Tempter's words; he had said that in the end he would see for himself that it was an astronomical observatory. The Tempter must have greatly overestimated him. There was nothing here he could recognize: no telescope, sextant, map of the stars, or brass model of the planetary system.

Instead, the circular wall was covered for the most part with unusual windows. They shone in a variety of colors, but it could not have been the light from outside because it was night. Some forms were moving on them, and he cautiously went up to one part of the wall to get a better look. They turned out to be yellow numbers that proceeded as far as the eye could see in horizontal rows against blue or red backgrounds, appearing at one end and disappearing at the other, although the device that was writing them was nowhere to be seen.

He would have stood there a long time staring at this sparkling display, whose meaning he had not even tried to penetrate, had it not been for the sound of quiet voices he suddenly heard behind him. He started in complete surprise. During his first moment of confusion, all he felt was the instinctive need to hide somewhere, but there was no time for that. When he turned around, just a few steps from him were two tall figures—a man and a woman—dressed in long white robes, heading his way, talking in hushed tones.

They had to see him; it was unavoidable since he was standing right there in front of them, paralyzed and bewildered. But they went straight past him, paying no attention to his conspicuous presence, as though he were completely invisible. He stood there for a long time, immobile, trying to get used to this impossibility, as his temples pounded fiercely.

The figures in white went up to one of the windows that was considerably larger than the others and was unlighted and started to touch some of the bumps that protruded under it. The window suddenly lit up, but it did not have the stream of numbers as on the others. It showed something that the prisoner could finally make sense of. The star field seemed far denser, brighter, and sharper, but basically did not differ from what he had seen through his small telescope.

But how could the picture in the window and the telescope be the same? What kind of window was that? The answer soon followed, but his readiness to believe took considerably more time. The two people continued to touch the bumps, and the scene slowly started to change. The change itself was clear to him, but he could not figure out how it was done. He would have achieved the same effect if he were slowly to raise his telescope: some stars would disappear under the lower edge, while others would appear above. But here the window did not move at all.

Then he heard something buzzing behind him. It was quite feeble, like the sound of a distant bee. He probably would not have turned around if he hadn't been compelled by the pins and needles at the back of his head—the tension of premonition. Something was going on behind his back, something big was moving.

The heavy, upright cylinder in the lower part of the slit in the dome slowly rose toward the highest point, although he could not see how it moved. It seemed to be doing so by itself, without the help of ropes and a winch.

He caught on to what was happening before the cylinder stopped at an angle of about seventy degrees. So, the Tempter had not overestimated him too much. In any case, it was only a matter of proportions here. Even though it was gigantic, the telescope had kept its original shape. What he could not understand was that the eyepiece had been moved. Instead of being in the only place it could be, at the bottom of the cylinder, it was on the wall like a big window that everyone could look at.

The picture on it stabilized just for a moment, and then a new change started. The stars began to flow over all the edges as though the telescope were rushing through the air at an unbelievable speed, although it was resting immobile. It penetrated more and more into the dark expanse, reaching for unattainable infinity.

The impression was intoxicating, delightful. And then, as if this were not enough, music echoed. The woman in white went for a moment to a smaller window and touched something. At the same moment, the crystal sounds of heavenly harmony reverberated from all sides. He could not see any musicians or instruments, he could not understand a thing, but he did not care. He was experiencing what one undergoes perhaps once in a lifetime: exaltation.

The two climaxes merged into one. One point in the middle of the picture started to get bigger, to expand. At first it was a star like the countless ones around it, then it was a circle, then a ring, and then finally it burst into a lacy flower that filled the entire window. The moment it opened its rosy, vaporous petals, the music streamed upward, greeting with an upsurge of joy the appearance of the yellow nucleus—the hidden eye of the Creator himself.

He was not filled with frustration when everything around him suddenly froze and became silent. He knew this would happen, that the watch cover had to open once again. The moment of the about-face was perfect. The epiphany had just taken place. Dared he hope for anything greater?

Return trips always seem shorter than departures. There were no more surprises and wonders to slow down time. Even though he felt awe as he watched the reverse sequence of what he had seen before—the disappearance of the dome, the return of the barred windows, the formation of doors and beds, the flickering of days and nights—his thoughts were elsewhere.

His confused thoughts that gradually formed a crucial question.

The end of the voyage came abruptly once again, just as when he had arrived in the future. At first, while his eyes were still blinded by the flashes, he could not make out anyone on the other side of the cell. Icy fingers of horror tightened around his chest. What if he wasn't there anymore? If he had only been playing with him? That would be just like the Tempter. Then he never would know...

"So?" came a gentle voice from the darkness.

He tried to muffle his sigh of relief, but such effort was futile in the murky silence of the night. "You said the observatory would be named after me, didn't you?" There was no time to beat around the bush; he had to get straight to the point.

"Yes."

"Why?"

"What do you mean?"

"Because of the discovery I made or because I was burned at the stake for not renouncing it?"

"For both one and the other, although considerably more for the act of sacrifice. You know, in the age you just visited, your discovery has only historical value. It has not been refuted, but it is secondary, insignificant, almost forgotten. As you have seen, things have advanced much farther. But your burning will not be forgotten."

From somewhere in the heart of the monastery came the sound of heavy footsteps. It was not just two guards. A larger group was walking through the corridors.

"Does that mean I have no choice?" asked the prisoner quickly. "If the observatory is named after me because I was burned at the stake, then it necessarily follows that there is no way I can avoid that fate. But I can still do it. I still have free will. They're coming. What if I say yes when they ask me to renounce my discovery? That would spare me from the stake but would change the future, wouldn't it? And the future cannot be changed; I saw it with my own eyes."

The steps ceased for a moment, and then in the distance echoed the harsh sound of a barred partition door being opened.

"That's right. You can't change what you saw. And you saw only that which is irrefutable, that which you cannot influence in any way. What you did not see, however, is whether the observatory is named after you."

The prisoner opened his mouth to say something, but no words emerged. His sight had returned in the meantime, so that now in the obscure light of dawn pouring in from the high window he could make out the contours of his visitor. His head was somehow elongated, as though he had something tall on top of it.

"No, I did not deceive you, if that's what you're thinking," he continued. "The observatory really will be named after you if you are burned at the stake. But if you are not, it will be named after someone else. One of your students, for example, who will be braver than you. There is no predetermination. Your free will determines what will happen. You will choose between a horrible death in flames and the penitent life of a royal astronomer under the wing of the Church, whose comfort will be disturbed only by the scorn of a handful of students and perhaps a guilty conscience: between satisfying your own conceit and the wise insight that it actually makes no difference after whom the observatory is named. I do not envy you. It is not an easy choice."

The rumbling steps stopped in front of the cell door, and a key was thrust into the large lock.

"You know what I will decide," said the prisoner hurriedly in a soft voice. It was more a statement than a question.

"I know," answered the gentle voice.

The rusty hinges screeched sharply, and into the small cell came first a large turnkey with a torch raised high and after him two Inquisition interrogators in the purple robes of the high priesthood. The soldier who entered last was also holding a torch. There was no more room inside, so the three remaining soldiers had to wait in the corridor.

In the smoky light the prisoner squinted hard at the figure on the bed across from him. The strange object on his head was some sort of cylindrical hat with a wide brim, and its slanted shadow completely hid the man's face.

He had not expected his visitor to stay there. Would he let the others see him? But no one paid any attention to him, as though he were not there, as though he were invisible. In other circumstances this would have confused the prisoner completely, but in the light of his recent experience he accepted it as quite natural.

"Lazar," said the first priest, addressing him in an official tone, "this is the last time you will be asked: do you renounce your heresy and penitently accept the teachings of our Holy Mother the Church?"

The prisoner did not take his eyes off the figure in black, but he had turned into a statue. He sat with head bowed, silent, just like an old man who had fallen asleep, with his white hands leaning on the top of his cane. He seemed indifferent, as if all this had nothing to do with him, as though he were not the least bit interested. The silence grew heavy with tension, with expectation.

And then at last, the royal astronomer turned slowly toward the inquisitors and gave his monosyllabic answer.

The Paleolinguist

I

The knock echoed loudly in the hollow silence, making her start.

She had not heard the steps approaching the door to her office. She must have dozed off again. Her head bowed, chin upon her chest, her round, wire-rimmed reading glasses had slipped to the tip of her nose. The book remained open in front of her on the desk in the lamplight, but she was still drowsy and could not remember its title right away. These catnaps were becoming more and more frequent, causing her to feel very ill at ease. Not because someone might find her in that unseemly position. She was not afraid of that; almost no one visited her anymore, not even her students, let alone her colleagues. She was an embarrassment to herself.

The knock came again. Brief and somehow reserved, hesitant. Certainly not as loud as it had seemed the first time. She looked around in confusion, wondering what time of day it was. The only window in her office looked onto the skylight, but this name was quite inappropriate since the narrow shaft that went through the middle of the building from the roof to the basement was filled only with gloom even on the sunniest days.

There was a simpler way to find out the time, but it would take her at least a few minutes to discover her wristwatch in the disordered multitude of large and small items that covered her desk. And she could not let the visitor wait that long, whomever he or she might be. Visitors were rare and therefore precious.

"Come in," she said. And then, since she thought she had said it too softly, she repeated in a louder voice: "Come in."

She did not recognize the person who appeared at the door. The neon lighting in the hallway illuminated him from behind, but even if the light had shone from in front of him, she would not have been able to discover very much without her other glasses that were also buried somewhere on the desk. The only thing she could conclude with certainty about the hazy outline was that he was a tall man in a dark cloak.

She pondered for a moment but could think of no one she knew who fit that description. That, however, still did not mean anything. She had learned with increasing certainty during the passing years that memory was a very unreliable support, particularly where the recent past was concerned. The more distant past was considerably sharper, which was rather apropos in view of her profession. But it made no difference: everything would become clear when the visitor started to speak. She had a hard time remembering faces, but she never forgot a voice, ever. This was probably the only department in which senility had kindly spared her from its humiliating veil.

"It's not easy to find you. You're completely hidden here in the basement." She had not heard this voice before. It sounded deep and drawn out, almost melodic. It would be impossible not to remember it, even without her aptitude.

"Oh, it makes no difference. When no one is looking for you, then it's all the same where you are. But are you certain that you're in the right place?"

"This is the office for paleolinguistics, isn't it?" It was more a statement than a question.

"Yes. Or rather what's left of it. In happier times we even had a brass plate that said so, but ever since we moved here, no one has taken the trouble to put it up. Maybe they're waiting for me to do it."

Continuing to stand in the doorway, the visitor contemplated the gloom of the rather small room. Three walls were covered with metal shelves, and the books and journals on them were more stacked, even thrown, than placed in an orderly fashion. A narrow vitrine rising to the low ceiling with its hot water pipes was on part of the fourth wall next to the window. It was full of tiny broken statues, pieces of pottery, and the remains of simple stone implements. These objects were also displayed without any order, often one on top of another as though the vitrine were a storage cabinet. Under the window next to the desk on a backless wooden chair covered with newspapers was a hot plate with a black kettle. Several used tea bags were lying on the newspaper like tropical fish that had died of asphyxiation.

"This is exactly as I imagined it," said the man at last.

"You imagined *this*?" she asked, bewildered.

"Yes, your office. Where you work."

She squinted, trying to focus her eyes better. "Is that supposed to be a compliment or a reproach?"

"A compliment, of course. What else could it be? I am an admirer of yours."

At first she did not know how to respond. She slowly took off her reading glasses and put them on the desk. When she finally spoke, her voice was critical. "If this is some sort of joke, then I must say it is rather out of place."

"Why do you think it's a joke?"

"I do not have admirers. I have never had any."

"But your work certainly deserves them."

She got up out of her armchair, numb from sitting so long, and started to rummage through the things on her desk in search of her other glasses. She hunted for several moments and when she couldn't find them, waved her hand in a gesture of angry dismissal, turned her blurry eyes toward the door and said in a voice that was more nervous than she intended, "Oh, come in, for heaven's sake. We can't talk while you're in the hallway."

He entered, closed the door behind him and then stopped, uncertain where he should sit. There was another armchair in front of the desk, but it held a load of tattered folders with a fairly large stone figure on the top; with the help of a considerable amount of imagination, it resembled a bulging female torso.

"Put that somewhere, on the floor, it makes no difference," she said, noticing that he did not know what to do.

He did it with utmost caution, as though holding some sort of relic in his hands. When he sat down, the springs on the armchair squeaked in protest.

Now he was closer to her and partially illuminated by the light from her table lamp, so that even without her other glasses she could make out certain details she had not noticed before. In his lap he laid his derby, his cane with its decorated top, and a pair of white gloves. She had never given much thought to how she dressed and did not pay attention to what other people wore, but she found this quite amazing. It was as though he had come out of a play set in olden times, she thought, smiling to herself.

The man just sat there without a word and looked at her. She soon began to fidget under his inquisitive stare. Unconsciously she started to fix her disheveled strands of gray hair as she thought over what to say to the stranger. Why had she asked him to come in? Admirer! As if she were so credulous or vain.

"So, you are interested in paleolinguistics?"

"Yes, very much so."

"Why?"

He did not answer right away. He started to draw his fingers slowly along the smooth edge of the derby in his hand. "An unusual question from someone who has devoted her entire life to that field," he said at last.

"Not at all unusual," she replied. "The very fact that I've squandered my whole life in paleolinguistics gives me the clear-cut right to ask you that."

"Do you think you have squandered your life?"

She stared at his blurry face, outside the lamplight. She could not guess his age. His voice was not a reliable indicator. Judging by it alone, the man could have been in his twenties or even his forties. For his sake, she hoped it was the former; it would be much easier for him to lose his illusions. If only she had been lucky enough to have some sense knocked into her at that age.

"Take a good look around you again. You are in a tiny basement room that was the janitor's storage before and will return to that function when I retire in several months. Since I am not able to take these things with me, the books and other artifacts will all be thrown away. Useless. And even if I took them, it would not make much difference. Everything would end up on the garbage heap after my death. There, that is the best measure of the success of a life devoted to paleolinguistics. So please listen to my advice: get interested in something else. Anything. Forget primeval language and the far-off past. Who is interested in that in the modern world? Don't ruin your future for no reason."

"The past and the future, yes," replied the visitor, lost in thought. He paused for a moment, and she thought a smile flickered on his face. But she could not be sure. "I think there are other measures that can be used to evaluate what you have achieved." He said it with determination, like a man who knows what he is talking about.

She looked at him inquisitively. "What, for example?"

"If it weren't for you, the department of paleolinguistics would never have been founded."

"Probably, but what has been the benefit of that? Do you know the greatest number of new students I have had all these years?"

He clearly did not understand this as a question and so did not reply. He did not even shrug his shoulders.

"Eight. And that was long ago; it's been almost a quarter of a century. The average has been three and a half students. And only two of them at most finish their studies. Sometimes not even one. But not because I was too strict. On the contrary, I was considered a very..." she stopped for a moment, looking for the right word, "helpful examiner, which gave me a bad reputation among my colleagues. The young people simply gave up, primarily because they were disappointed, even though I did all I could to stimulate their interest not only in the technical aspects of the origin of language but also in a considerably less tedious subject: early human communities. They are inseparable, in any case. But nothing seemed to work. I never understood what they actually expected when they decided to major in paleolinguistics. No one made them choose it."

"You cannot blame yourself for the students' poor response. You said yourself that we live in a time that is not particularly predisposed toward the past."

She squinted at him briefly, and then continued to follow her line of thinking, paying no attention to his comforting words.

"In the last four years, no one has signed up in my department. How can you keep your position as lecturer if you have no one to lecture to? Only if the administration is sympathetic toward you. They didn't have to do it. They probably wouldn't have if it weren't for my age. I stayed here just because the dean was considerate enough to support me, although it would have been natural to fire me. He knew that at my age I have nowhere to go. I knew that myself, so I swallowed my pride and let them put me in this cubbyhole. Don't look a gift horse in the mouth, particularly not when the gift is given out of pity. What else could I have done, anyway?"

She stopped talking, wondering why she was telling all this to a stranger. She was only putting them both in an awkward situation. But the matter concerned him, too. He had come there with an idealized notion of paleolinguistics, hadn't he? Would it be fair to let him leave without seeing its other side? Certainly not. In any case, she had not had the opportunity to talk to someone for a long time, to pour out her grievances. There were no more students, and her colleagues avoided her more-or-less openly.

"Now I'm on sabbatical. That was the last chance for me to reach retirement age in this position. I was given a leave of absence quite easily. It was actually a gift. I didn't even have to present any sort of research plan, as is customary. No project that I would work on. No one even asked. No one expects anything from me anymore."

"But you have done so much already. You wrote several fundamental works on paleolinguistics. Isn't that more than enough?"

Her blurry eyes started to wander over the multitude of objects covering the desk in front of her. Had she known she would have a visitor, she would have tidied up the office a bit. Actually, she had been reproaching herself for some time for the clutter surrounding her, but she could never make up her mind to do anything about it. There was no incentive. What was the purpose, since she would be leaving there in a few months? But then, couldn't that be expanded to all of life itself? Why make any effort at all when everything was transient? She used to know the answer, it had seemed obvious and irrefutable, but with the passage of time it had become hazier and darker.

"Would you like some tea?"

He did not answer right away. He seemed to hesitate. "No, thank you," he said at last.

It was only then that she realized there was just one tea cup. Had the man accepted her offer, she would have had to do without tea, something that would not have been easy for her. She had become a complete addict. Several years ago, when the doctor had advised her to stop drinking coffee because of high blood pressure, she had switched to tea, primarily to appease her habit of constantly sipping something hot. When she reached seven cups a day, she realized she had overdone it, but it was too late by then.

"I would. Do you mind?"

"Not at all."

She hobbled over to the small, cracked ceramic sink that stood next to the window, picking up the kettle on her way. Although her vision was very poor without her glasses, she did not need them to make tea. She had gone through this sequence of simple motions so many times that she could have managed in total darkness.

"There is nothing truly fundamental in my works," she said in a hushed voice, after plugging in the hot plate and returning to her armchair. "It is all just an educated guess, at best."

For a few moments all that was heard was the sound of water leaking in thin streams from several spots on the cracked exterior of the kettle, evaporating when they hit the red-hot plate.

"What do you mean?" asked the visitor at length.

"Do you know the first thing I told my students? So they knew right from the start what they were involved in. Paleolinguistics is not an exact science. It cannot be, since, in the strictest sense of that term, the subject of study is missing. Primeval language has been dead for a very long time. We have no direct evidence of it. And even the indirect evidence is quite scanty. All we do is make more-or-less questionable reconstructions. We try to recompose a mosaic whose original appearance is unknown, and we are not even certain that we are using the right stones."

"But didn't you convincingly show that living languages and the dead ones that have been preserved both contain traces of primeval language? Which is natural, in any case. They all arose from it, didn't they?"

"Convincingly, yes. Perhaps. There is one person, however, whom I have never managed to convince of this completely. The only one I really care about."

"Who is that?"

"Myself, of course."

The kettle suddenly whistled. She got up slowly, unplugged the hot plate, took a small tea bag out of a half-empty yellow box on the desk; lifted the kettle's little lid, removing her hand quickly so the steam would not burn her, waited for the cloud rushing out to disperse, and dropped the tea bag into the boiling water.

"They say you shouldn't put the tea bag in right away. If the water is too hot, it kills the aroma of the tea. But I don't have the patience to wait."

"You are unfair to yourself. You must not doubt your whole life's work. Just think of the enormous effort you have made."

"What else can I do? Resort to self-delusion? Repeat to myself that it can't be all in vain since I made such a tremendous effort? But effort itself is by no means a guarantee of success. There is something, however, that is even worse than doubt. The hardest thing for me is that the doubt can never be removed: there is no way to know how close I came to primeval language. But there's no one to blame for that. I knew from the beginning that was the main short-coming of the field I had chosen."

"Except if you were to go back into the past."

She smiled at him briefly. "Yes, except if I were to go back into the past. I know many people who would sell their soul to the devil without the slightest hesitation for such an opportunity. All kinds of historians. People like me, obsessed with long-ago times. But either the soul is not enough payment or the devil himself is not that powerful. Probably the latter. Unfortunately, there's no going back into the past."

"If there were, would you accept the devil's offer?"

She looked at him without speaking for a time and then got up to wash out her cup and pour the tea. When she returned to the desk, the newspaper that covered the chair with the hot plate had a new fish steaming in agony.

"I don't think the devil would choose me. What use would he have for such a poor, worn-out soul as mine?"

"Maybe he wouldn't even ask for your soul."

"Oh, don't be naive. The devil isn't generous. You don't get something for nothing from him."

"I agree. He always collects payment for his services. But there are other rewards in addition to the soul that he might find more attractive."

"What else could he expect from me in return?"

"The devil is a sadist above all. He enjoys people's suffering. If he helped you return to the past, he would be putting you in twofold torment."

She greedily took a sip of tea. She knew it was still hot, that it would burn the sensitive lining of her mouth, but the addict in her had run out of patience again. Conversation with this stranger had become rather pointless, even though he amused her in some odd way.

"Twofold?" she repeated inquiringly.

"Yes. Imagine that you go back in time and there, on the spot, you reliably establish how things were. What would you do with that knowledge?"

"Well, I don't know. Publish it, probably."

"But you are a scientist. Wouldn't you ruin your credibility by citing that your knowledge stems from a trip into the past arranged by the devil? They would proclaim you a charlatan at best. At worst you'd end up in an insane asylum."

Before she replied, she took another long sip of hot tea. The cup was already half empty.

"Then I wouldn't publish it. But the devil would still have no reason to rejoice. I told you that I only care about convincing one person. And for her sake it would not be necessary to publish anything. She would be convinced without it, by firsthand experience." She stopped a moment, smiling again. "Let me hear what other trap the devil has prepared for me."

"What is the fundamental assumption of your field, that is, all fields that study the past?" The visitor had not acquired her facetious tone. His voice was as serious as before, and she thought it quite pleasant. Dignified. Too bad she had not found her glasses. A man with such a voice simply had to have an agreeable face.

"The immutability of what has happened, if that is what you had in mind."

"That's right, the past cannot be changed. That fact would be jeopardized, however, if someone from the future appeared in the past. The devil's services would desecrate something that is older and must remain inviolable. What would be the use of learning firsthand about the past if it were no longer final?"

"Why do you think that a visitor from the future would destroy the past? If he were a scientist—and we're talking about that kind of time traveler, aren't we?—it would not be in his interest. On the contrary, he would have every reason to remain an inconspicuous observer."

"Yes, he would have every reason. But would that be enough? There would be enormous temptation to influence the course of events. Take, for example, a historian who goes back to some turning point in history. If he remains simply an observer, events will take their well-known course resulting in the death of a large number of innocent people. On the other hand, it could all be avoided by his involvement. In that case, which would prevail inside him: the dispassionate scientist or the man who realizes that if he does not take any measures, his conscience will be burdened with unbearable guilt? It would not be an easy choice, and this would give the devil great satisfaction."

She stared for a moment at the bottom of the empty cup in front of her before answering.

"Not every traveler to the past would necessarily come up against such a difficult choice. There are peaceful times, without turning points. For example, if I went back to the period I studied, I could be an impartial observer without any encumbrance because nothing would drive me to get involved in the course of events, to change the future. Historically speaking, it was a completely innocent age. I'm afraid the devil would not get his due."

"There is no innocent age." He said it softer than before, as though it were confidential, secret. "Have you heard of the butterfly effect?"

She had heard of it but could by no means remember what it was. Even if her memory had been in better shape, it would quite likely have slipped her mind. She had never fancied such innovations. Her science was classical, more elementary. To avoid answering, she got up to pour a new cup of tea, and he waited for her to return to the desk.

"A butterfly suddenly starts to fly, urged by who knows what, quite unaware of the fact that this movement might start a chain of events whose far-off final link is a storm of continental proportions on the other side of the world. The flutter of tiny wings sets the chaos equation in motion, whose solution can be completely disproportionate to this infinitesimal movement. A tiny cause sometimes leads to enormous effects."

"Yes, I know about that," she replied, "only I don't see what that has to do with what we were talking about."

"Regardless of how firmly you are resolved not to change the past, what happens does not depend on you alone. Quite unintentionally, by your very presence, you might bring about the butterfly effect. Maybe even literally. Imagine that your sudden appearance there disturbs a butterfly that has been idly perched on a flower. Frightened, it suddenly takes flight, and several days later, far from there, someone dies in a storm who was not supposed to die at all, someone who is the starting point of an inverse pyramid of history. You might be convinced that this outcome is highly unlikely, but the devil, as an experienced gambler, would not hold back from accepting the wager. It would actually be a safe bet. Chaos is his kingdom, when it comes right down to it."

"But if this is how things stand, if the devil can't lose, what's stopping him from coming with his offer? He hasn't visited me, or anyone else in my field as far as I know. And it's among us he would find the most prominent victims."

She expected an answer from the other side of the desk, but none was forthcoming. As the silence in the gloomy basement room deepened, distant unintelligible sounds from the upper levels could be heard.

"So, we are back to where we started," she said, finally breaking the silence. "Going into the past is clearly not within the devil's power."

"Maybe it is," said the melodic voice in return, "but in such a way that it would not give him the reward he wants if he were to offer it to someone. That is why there is no offer."

Now it was her turn to remain silent. She peered in bewilderment at the foggy figure across from her.

"If there were neither the temptation nor the opportunity to change the past, then the returnee would feel none of the torment that would suffice the devil as payment."

"But is that possible? Doesn't it follow from your story about the butterfly that the very act of stepping into the past would inevitably change it?"

"Yes, it does follow, but only if one went physically into the past. And it does not have to be that way."

She raised her cup to her lips, but the tea was already lukewarm. This was not the way she liked it; it was tasteless. She found a bit of empty space on the desk and put the cup there.

"Then how would it be?"

"What do you do when you watch a documentary film?" said the visitor, answering her question with one of his own. "You go into the past without the opportunity of changing anything. Film editors do have a few tricks at their disposal, but that doesn't count: that would be falsifying the past and not truly changing it. The viewer of a documentary film is in the position of the ideal unbiased observer: he can in no way influence the past."

"Yes, but that is only true for more recent history. It really is possible to return to the past that way. At least partially. The filmed version enchants us with its images and sound, but reality is something richer. But let's put that aside. I must remind you that, unfortunately, no documentary films have been made about the age that interests me."

If he noticed the irony in her voice, it was not revealed by any change of tone. "Of course not. I was not even thinking of such a return to the past. It is, as you say, quite incomplete. But the comparison with films is rather convenient. Imagine such a film about the past that would act upon all your senses, not just sight and hearing. A film in which you would feel exactly the same way you do in reality, except that you could not take part in it, change it. You would have the role of an infinitely empowered viewer who sees, hears, and feels everything, yet remains invisible and inaudible, unobserved."

She blinked. "That sounds like a ghost to me."

The visitor gave a brief laugh, resonant and clear. "Yes, the viewer would be like a ghost, for all practical purposes."

"That's all very well, but there are no such films about the past. None have been made."

She thought there would be some comment from the dusty armchair, but silence greeted her once again. She closed her eyes for a moment and rubbed the bridge of her nose with thumb and forefinger, thinking that it was time to bring the conversation to a close. What else was there to say? They had reached the topic of ghosts, hadn't they? Even though she was happy that someone had visited her, she was now feeling tired. A person should not be overly indulgent toward admirers.

"I wonder what time it is. My watch is lost somewhere in this mess on the desk. I can't wear it on my wrist all the time—it chafes—and then I can never remember where I put it."

She assumed that he would look at his left wrist, but he reached under his cloak and took out a pocket watch. A dull golden reflection danced about it, and she thought how strange it was. She could not remember the last time she had seen a watch like that. Had they come back into fashion? She knew nothing about fashion, but then how could she since she never went anywhere, never saw anyone, and passed the entire day between these four basement walls?

He handed her the watch. She took it impassively, simply because he had offered it. It was only when she had it in her hand that she wondered why he had not simply told her the time instead of letting her find out for herself.

She brought the object close to her eyes, so she could see without her glasses, but did not open the lid right away because her attention was attracted to the engraving on its bulging surface. An ornate letter E, with a series of decorative loops at its ends, just like the initial from some old-fashioned manuscript. Quite unusual, was the thought that flashed through her mind. E as in Eva. Like it was meant for me.

And then she moved the little catch and the lid jumped up.

II

There were no hands. There was no face. Just a bright circle that contained some kind of image. The image was not quite steady but trembled as though alive. Confused, she brought the watch a little closer to get a better look, but when she wanted to stop, it kept on coming closer all by itself, without her influence. The casing started to get bigger, like a round fissure in reality that quickly expanded before her, its brilliance crowding out the gloom of the basement room, until it had pushed it all the way over the edge of the world.

She was blinded at first. Her pupils were accustomed to the poor light in the office and needed some time to adjust to the bright midday sun. But the rest of her senses immediately began to absorb the rich impressions of her new surroundings. She was struck by the unknown smells of wild vegetation, dense and abundant, prickling and stinging her nostrils as though someone had thrown a handful of pollen into her face. Her ears were filled with the undulating sound of tall, brittle blades of grass and the buzzing and humming of a multitude of insects engrossed in their ritual dances. The breeze reached her skin in uneven gusts, stroking her face and hands with the softest touch.

She knew what she would see even before her eyesight returned, but there was still no lack of surprise. She was in the middle of a field that stretched all around in gentle folds as far as the eye could see, but what her senses of smell, hearing, and touch could not tell her was that countless butterflies covered the expanse around her like some flickering, brightly colored rug. They were flying low over the ground cover or resting on it, completely devoted to their harmless business which, as she had recently learned, could result in unforeseeable disaster.

She froze at that thought. What if her appearance upset them? What if they suddenly started to fly, thereby disturbing something that should not be disturbed? She stopped breathing when a butterfly left a purple flower with large petals and zigzagged toward her, lazily fluttering its spotted wings. When it approached her face she instinctively closed her eyes, helplessly expecting it to fly into her at any moment. But no crash occurred. When she opened her eyes, the butterfly was gone. She first thought that it must have turned at the last instant. The other possibility was so unbelievable that she simply refused to accept it.

But soon afterward, when a somewhat stronger gust of wind raised an excited cloud of butterflies, she nonetheless had to accept the impossible. They flew through her as though she were not there, as though she were made of some airy substance, transparent, unreal, nonexistent.

At first she just stood there motionless, utterly confused, and watched the cloud stream through her body. She felt this rising tide like a weak sting, like light goose pimples flowing on the surface of her skin. The cloud had already thinned out when she finally emerged from her paralysis and extended her hand toward the last butterflies. She could touch them in flight. The touch was irrefutable, although one-sided: the tiny wings yielded unfeelingly to her invisible fingers in their multicolored fluttering.

She remained undecided for a time after the last butterfly had gone. Serious, distressing questions were welling up from part of her consciousness, but she quickly smothered the tiresome voice that only spoiled the magic. What difference did it make that it was impossible when it seemed so dreamy, so intoxicating?

Nevertheless, one question had to be answered. What next? She could stay there some more—quite a lot more, actually—surrendering to the fragrances emanating from the ground, the salutary warmth pouring down from the firmament, the caressing wind that wakened inside her long-hidden joys. But not for an eternity. Even the Elysian Fields inevitably lose their charms.

She had no reason to go in any particular direction, so she simply moved straight ahead. She did so unconsciously, taking a step forward, but instead of her foot landing on the grass again as it should have done, it stayed in the air.

She did not realize right away that she was flying. At first she thought she'd lost her balance and would fall, but she never did. She remained in midair, unsupported, bewildered because she had always been afraid of heights, although she was barely at knee level. She wished in panic to go down, and the very next moment she was resting on the ground again.

Some time passed before she mustered the courage to move once more. She thought she must look like a child, trying to take its first awkward steps. This time there was no need to step forward. All she had to do was will it: she wanted to fly—and the same instant she was in the air again, infinitely light, incorporeal.

She first took a horizontal birdlike position, extending her arms like wings, but quickly realized that this was not necessary. Undignified, in fact. Owing to her years, it was much more becoming for her to adopt the same position in the air as she did on the ground, so she straightened up with her arms crossed on her chest, as though perched on an invisible pedestal.

Fear faded and gave way to fascination. The experience of unhindered flight was thrilling, giddy. First she streaked high up until she reached the fluffy substance of a small cloud, and then, barely resisting the urge to scream with excitement, she started back down, enjoying the sight of the green carpet approaching at lightning speed. She stopped right above it effortlessly, without disturbing the swarm of buzzing insects quarreling over a cluster of red and yellow flowers.

When she soared to the bottom of the heavens again, she caught sight of something she had missed her first time up there. Her surprise was actually twofold, and she stopped suddenly in the middle of nothingness. When she saw a thin column of smoke rising on the distant horizon, it flashed through her mind that this should not be possible: she did not have her glasses with her. They had been left behind in the office, somewhere in the disorder on her desk. But it seemed that in this new form they were not necessary; she could see the spiraling signal of someone's fire clearly enough without them.

She hurried in that direction like an eagle that has spotted its prey, driven by impatience and foreboding. The suppressed questions started to surge to the surface once again. If she was truly where she suspected, although all this was beyond reason, of course, then she had lighted on her destination.

The tribe was small—she counted only twelve members. Next to the fire were two old women, an old man, and four children of different ages. The other five adults—how stunted they were!—were dispersed in a broad circle around this temporary habitat. They were engaged in what people of that early age spent most of their time doing: painstakingly collecting food—different berries, roots, shriveled fruit, small rodents.

She descended, not close to the fire, but a bit farther off. She could feel her heart thudding in her immaterial chest. The voices of the old people and children were muffled and indistinct, but that was what she wanted. She was not yet ready. When she was, she would go among them—a ghost who would know as soon as she heard their first words whether or not her former life had meaning.

She wondered what price she would have to pay for this unique privilege. It certainly could not be the assurance that she would not return from here. Even if she had that impossible watch, what was there to go back to? Lonely drudgery in a dark basement cubbyhole? The humiliation brought by neglect and old age? The implacable doubts that would follow her maliciously to the end? No, staying here would be a reward and not a punishment. So what would it be?

The answer came with the wind. The current of air brought to her insensate nostrils the hot smell of steam from the sooty earthen vessel in which water was boiling over the fire. The old women were cleaning some dried herbs, getting them ready for the pot, chatting idly, just as would be done in the countless centuries to follow.

Tea, of course!

An inaudible scream was wrenched from the addict. She felt neither hunger nor thirst, as was quite natural in this state. But the longing for a hot cup of tea that suddenly flooded her was something far beyond a physical need. The delusive impression that the familiar tonic was flowing through the inside of her mouth, the promise that her overpowering need would soon be satisfied, had the same effect as genuine agony.

As despair filled her, she thought that she would not have accepted had she known the price she must pay. But she had not actually been given the choice. All right, then, she concluded, getting hold of herself, there's no turning back. The price has been paid, even though unwillingly. All that was left was to take what was hers in return.

And she headed toward the fire to meet the voices of the primeval language that would tell her the simple truth.

The Watchmaker

I

The clocks struck six p.m. simultaneously, just as they should in a reputable watchmaker's shop. The old man's trained ears had been carefully monitoring this sound, and they could not detect any divergence: not a single one of the four clocks adorning the walls of the rather small, ground-floor premises was either early or late. This was the only harmony that linked them, however, for what followed afterward was total discordance.

The grandfather clock, with its pendulum in a casing of worn mahogany and door of thick, etched glass, grumbled in a deep, solemn bass, like a mustachioed sergeant grenadier giving orders at a parade. The brass dwarf hit his worn hammer on the hanging bronze rod, creating a clear, sharp sound resembling the echo of distant bells. The call of the wooden cuckoo rushing out the round opening of the gaily colored alpine house had lost its original rapture long ago, becoming harsh and piercing. Finally, the chipped pair of ceramic dancers in ballroom attire nimbly started to turn on the small circular podium at the first bars of an old-fashioned waltz.

Although they began all at once, the sounds that struck the hour did not end at the same time. First the cuckoo went silent, suddenly, like a death rattle; it seemed almost as if someone with delicate nerves or no ear for its tired singing had ungraciously wrung its neck. The waltz and the ringing lasted about the same time, competing to the final note for futile advantage. The drawn-out tones of the grandfather clock filled the shop the longest; by virtue of its very size it was natural for it to have the last word.

When the final grumble of the grenadier's bass had died out, the old man reached adroitly for a small pocket on the left-hand side of his vest. He took out a gold-plated pocket watch with a thick chain, raised the lid—which had TO J. FROM M. engraved on the inside in large, ornate letters—and briefly nodded, satisfied that it was truly six o'clock. This was not an expression of distrust toward the other clocks which had just informed him of the same fact quite loudly and precisely. For more than a quarter of a century he had carried out this ritual every evening before he closed the shop and went home, as a sign of respect for a special memory. And a grief.

But he was not fated to spend that evening in the usual way: closing the door to the shop, taking the short walk along the usually empty street to the small, excessively neat attic apartment where no one waited for him, preparing a simple and for the most part tasteless meal that would probably satisfy only a bachelor or a single person, and going to bed. Sleep would rarely bring him refreshment or oblivion; it mainly gave him restless dreams that returned him to the past. He could not leave the past, not even in his dreams.

He had just put the heavy watch back into his vest pocket and was about to pull the little short chain with its silver ring to turn off the lamp with the green shade on his workbench behind the counter when the door opened suddenly, jangling the cluster of bells hanging above. Although mild compared to the discordant choir of the wall clocks, the unexpected sound of these signal bells made him start. He rarely had customers in his shop this late.

He looked up, but all he could make out in the gloom was the silhouette of a tall man against the dull glow of the streetlight. The man was wearing a hat, probably a derby, and rather a long cloak, and in his right hand was a cane. He stopped next to the door without going up to the counter, as though hesitating for some reason.

The old man pushed his round, metal-framed work glasses halfway down his nose and asked, with an effort to sound obliging, "May I help you, sir?"

The man did not reply at once. He looked around the shop as though wanting to make sure that the two of them were alone. His eyes rested a bit longer on the grandfather clock; half of the pendulum's path was in shadow, and the circular base flickered in the other half as it reflected the muted light from outside.

The late visitor finally put his cane under his arm and stepped resolutely toward the counter, at the same time removing something from an inside pocket. When he reached him, the old man saw that he was wearing white leather gloves; he had long, slender fingers like a pianist's. His right fist was closed, and he placed it palm up on the felt-covered counter. Illuminated by the rim of light from the work lamp, its whiteness looked unnaturally bright compared to the green background and the darkness around them. The watchmaker suddenly had the impression that the man before him was a magician who was about to pull a sleight of hand.

The trick, however, did not take place, for when his hand opened, it contained quite an exemplary object: a pocket watch. The old man returned his glasses to the bridge of his nose and leaned over it to take a better look. Up until then, he had been convinced that all he needed was one look at a watch in order to recognize not only the brand but also the type and even the year it was made. He had spent almost four decades working exclusively with watches. He knew them inside out, one might say. Particularly pocket watches; he was a real expert where they were concerned. He knew each little spring, gear, screw, and nut. Every little hand and face.

But here he had a surprise in store. One look was not enough. He had certainly never seen this type before. The old man knit his brow in disbelief and leaned a bit closer. He was filled with the powerful urge to take the watch from the white palm, to finger it, open it, but manners prevented him doing so. He continued to look at it, putting his eager hands behind his back. He strove hard to find some detail he could recognize, but all his trained eye could ascertain was that the watch was exquisitely made. There was no doubt about that: it was the creation of a true master of his trade—an expert he had never heard of.

Shaking his head briefly, he straightened up and looked at the visitor inquiringly. The man's face was still in the darkness under the hat brim, so the watchmaker could not make it out. Suddenly he felt a mild prickling sensation at the base of his neck, the bristling of sparse white hairs. There was something unreal about the tall figure in front of him, something that filled him with agitation and unease.

This impression did not pass when the visitor finally spoke.

"I would like you to have a look at this watch," he said in a hoarse, dignified voice which did not need to be raised even when giving an order. A foreigner, concluded the watchmaker. Although he made an effort to pronounce the words properly, his accent gave him away as well as a certain drawl, although not one common in travelers from the north, who were the most frequent strangers in this area. It was impossible to say where he was from.

"Certainly, sir, certainly," he replied. "What is your complaint, sir? I mean, what is wrong with your watch? It is obviously quite expensive, although..."

He opened his mouth to admit that he had never seen one like it before, but he held back at the last moment, fearing this might stop the visitor from leaving his watch with him. He certainly had to have the chance to examine it in greater detail.

"I have no complaints," said the stranger, interrupting him. "The watch is fine. But all the same, I think it would be a good idea for you to have a look at it."

"Most certainly, sir. You are quite right. A bit of precaution would certainly do no harm. On the contrary, never enough caution. You were very wise to bring your watch to be looked at. Even the best watches need regular maintenance. People do not bear that in mind, actually, they are negligent for the most part, not only toward objects, unfortunately; many misfortunes would be avoided if precautionary measures were taken..."

"There are no precautions that can thwart chance." The man said this in an even voice, as though saying something obvious, even banal. The watchmaker squinted toward the invisible face; although the statement sounded like a general principle, there was something in the stranger's tone that gave it the weight and credentials of personal experience.

"Yes, indeed. Of course. You understand things perfectly, sir. Chance, yes. Something you cannot influence regardless of how hard you try. For a watchmaker that is the effect of dust. I have yet to see a watch without dust, and countless numbers have gone through my hands in my many years of work. You can protect a watch however you want, even close it hermetically, but nothing helps. Dust will find a way inside, and one particle is enough—one single, solitary particle—to jeopardize the fine mechanism. You have no idea, sir, what a nightmare dust is for watchmakers."

"Yes, a particle of dust," repeated the visitor, drawing out his words, lost in thought. "The flutter of butterfly wings..."

The old man's eyes became suspicious. What was that supposed to mean? What "flutter"? Maybe he wanted to say something else but expressed himself awkwardly in a foreign language—although he seemed to speak it well, at least fluently and correctly, if not without an accent. Or maybe he was some kind of crank, an eccentric? The old man was not prejudiced against foreigners and considered the stories that could be heard about their peculiarities, even abnormalities, to be exaggerated for the most part. But you never knew. There were quacks everywhere, in any case. Not even this neighborhood had been spared.

He had the impression that some sort of reply was expected from him, but did not know what to say. Really, "butterfly wings" ... What could he say about them and still be nice, polite? He was saved from the awkward situation by a carriage that suddenly passed by in the street. The rapid thud of horses' hooves caused the plated wheels to produce a sharp rattle as they rolled over the cobblestones. The visitor seemed to flinch a bit at this noise, turning toward the entrance. But the carriage passed in a flash, and the fading echo of its passage was quickly absorbed by the heavy silence of the evening.

"Yes," said the watchmaker when the stranger turned his unseen face toward him once again, "you are completely right. There is no way to fight against chance."

"Oh, that's not what I said. I only said that you cannot thwart it, prevent it. But that does not mean that you cannot fight against it."

The old man involuntarily swallowed the lump in his throat. "Please forgive me, sir, but I'm afraid that I don't understand you very well," he replied timidly.

Before he answered, the visitor finally put the pocket watch on the felt-covered counter, as though for some reason he had concluded just at that instant that he could safely let the watchmaker take his valuable timepiece. When the white glove withdrew from the lamplight, the old man had the impression that a bright trace remained behind it for a few moments. With his free hand, the foreigner skillfully took the cane from under his arm, turned slowly on his heel and pointed at the clocks on the four walls with it.

"It is all a matter of time, you see," he said at last, after making a full circle and returning to face the watchmaker again. His voice took on that flat quality once more that spoke of reliable knowledge, his own experience.

The old man simply nodded, without a word, as though this statement explained everything. One had to be careful with eccentrics; it was not advisable to contradict them.

"What makes chance so powerful? The fact that you can't foresee it. If you knew exactly which particle of dust would ruin the watch mechanism, you could remove it in time. But you can't know that until the malfunction occurs, of course."

"Of course," repeated the watchmaker like an echo, with another nod.

"Cause and effect," continued the visitor. "The particle only becomes a cause when the effect takes place—the malfunction. Never beforehand. That is why alleged clairvoyance and similar illusionary sleights of hand have no meaning. The future cannot be foretold because then one would be able to change it. And if you changed it, then it would no longer be the predicted future. You cannot prophesy: this particle is the cause of the future malfunction—and then remove it, because then there would be no malfunction, and your prophecy would have no value, either. No, the consequences must happen in any case. And they do take place. You yourself said that you have never seen a watch without dust inside. And you undertook detailed precautionary measures, everything that was within your power, to prevent it."

"Oh, I did, I did, most assuredly. You can be certain of that, sir. I hope I am not being immodest when I say that this watch repair shop has an excellent reputation for industriousness. You will see this for yourself, sir, I hope. We leave nothing to chance here..."

The old man stopped, biting his tongue; it was only after he had said this last sentence that he realized the expression he used might sound inappropriate, given the topic under discussion. But since the visitor did not react, he quickly continued.

"But, if you will forgive me my poor perception, sir, I cannot see how it is possible to fight against chance—your very words, sir—if the effects, the consequences, must take place?"

The foreigner did not answer at once. Led by some obscure impulse, he threw his cane a short distance into the air, then as it fell caught it adeptly near the upper end with his thumb and forefinger and started to swing the lower part as if it were a pendulum. It was only then that the old man noted in the gentle, milky gleam that the top of the cane was the stylized figure of an hourglass. Most likely made of ivory, he concluded. The man was without doubt quite wealthy. Perhaps only people like that could allow themselves the luxury of being eccentric.

"It's all a matter of time, as I said," he announced again at length, continuing to swing his wooden pendulum. "You truly cannot influence the cause *before* the effect, but there is another possibility—perhaps you can do so *after* the effect takes place."

The old man squinted again over the metal rim of his glasses. Watchmakers are like doctors, he thought, self-pityingly and comfortingly: they do not enjoy the privilege of choosing their clients. How would it look if a doctor refused to treat a patient simply because he had strange convictions? Should he now refuse to serve this obviously wealthy quack with a very unusual watch just because of his peculiar ideas? That would be quite against professional ethics, not to mention courtesy. And after all, there was the fee to think of.

"Oh," replied the old man briefly, trying not to sound too surprised.

"Yes," continued the visitor, "although extremely unusual, the idea is actually simple. Going into the past. Going upstream on the river of time, to put it picturesquely. If you returned to the past, you would be able to remove the cause and thereby the effect as well."

"Of course," agreed the watchmaker without hesitation. "Quite simple, as you said, sir... Going back into the past and removing the cause... Nothing easier, so to speak. No cause, no effect. You explained that quite well, sir, quite concisely..."

The stranger did not reply for several moments, and the old man had the unpleasant impression that the unseen eyes were gazing at him in suspicion from under the hat brim. Did I say something I shouldn't have? he wondered. Maybe I shouldn't have said anything. A man doesn't know how to talk to such people.

"It is not quite as simple as you might think." The visitor's voice seemed to carry a touch of reproach. "Here's an example: imagine that you go back to the past and accidentally cause the death of one of your parents—before you were conceived. That would mean that you were never born and could therefore never go into the past and prevent your own conception. And if you were nonetheless born and then you went back to the past...and so on. *Reductio ad absurdum*. A paradox."

The old man stared fixedly at the dark figure before him, suddenly feeling sweat break on the palms of his hands. What was he talking about—causing the death of one of my parents? How could he think of something like that? Was that the sort of thing a gentleman talked to a stranger about, even if he was an eccentric? But what if this person before him was not some rich eccentric, but a madman who had escaped from a foreign asylum for the mentally ill, who would rob and maybe even kill someone? Where did he get those fancy clothes, expensive watch and ivory-tipped cane, anyway? Does he intend to attack me? What should I do? How were you supposed to act toward a dangerous lunatic, anyway? Humor him, flatter him? I must not let him know that I realize he is crazy. But they say that madmen can be very bright... If only the ceiling light was on—damn the penny-pinching of the elderly!

"No, there is no solution to the paradox, at least not if you hold to the normal view of time—as a unique river. What has happened cannot be changed at all. The flow of time is like granite in which events are permanently chiseled. Both causes and effects. It is not a palimpsest that you can erase and write on again as many times as you want."

Another short pause ensued, and then the foreigner suddenly stopped the monotonous swinging of his cane. He held it in the hanging position for a moment, as though uncertain what to do with it next, and then with a sharp movement put it under his arm again. All that remained sticking out at the front was the figure of the hourglass—a milky spot before a dark background.

"But what if there were not just one time flow, one inscription in granite? If there were several flows—countless, actually? Imagine time not as a single river but as an enormous tree with countless branches, countless forks. Forks appear on those places where you change the past. One branch is the original flow in which a cause produced an effect; that is final—it must remain unchanged, chiseled—but from the other branch both the cause and the effect are removed."

The visitor stopped, as though wanting to check the impression his words had made. The old man was still staring at him fixedly, his mouth half open. In the sudden silence, the muted ticking of the wall clocks rose by several octaves.

"And you exist on both forks, in both versions, if we can put it that way. You have a sort of double—more than that, actually—whose course of life differs from yours in some respect. In an essential respect, perhaps. He could be spared the effects of an unpleasant, tragic accident, for example."

The visitor fell silent and the old man started to fidget, feeling that he should say something in reply. However, for several long moments he couldn't think of anything.

"Truly quite clever," he said at last, making an effort to keep his voice from trembling. "What an unusual notion! You have figured out something quite brilliant, sir. A tree and then a fork, and a double! Very picturesque, striking, no doubt about it. Something like that certainly would never have crossed my mind."

"Strange. And one would say that you have had both an opportunity and a motive to think about that."

"What are you thinking of, sir? I'm afraid I don't quite understand."

The visitor took the cane in his right hand again and described a rather large arc in front of him.

"Isn't this an opportunity? Look around yourself. You have spent your entire life in the midst of clocks. You are surrounded by chronometers. You are in the very center of time, I might say, in a very privileged position. I cannot believe that in all these past years you have never wondered about the nature of time, how it works, about the peculiarities linked to it. Who else if not you?"

"I am afraid you highly overestimate me, sir. I am just an ordinary watch-maker. Industrious, that is true, yes, and probably good, too, at least that is what they say, but nothing more than an artisan. For me, sir, and please don't hold it against me, time flows as it flows, and if a clock does not measure it as it should, I repair it. I can do that. And that is all. Clocks are here to measure time properly, aren't they?"

"Yes, that's true, but what about the motive?"

"Motive, sir?"

The stranger did not continue right away. The watchmaker could almost feel the piercing look of the eyes in the shadow.

"Nothing in your life has ever made you want to go back to the past and change something there? Remove some unforeseen cause that led to adverse effects? Cancel the consequences of some mischance that befell you or someone particularly close to you, someone dear? Has there ever been a man who has never had such a desire?"

Who is this? wondered the watchmaker in fear, feeling suddenly squeezed, as if in a trap. Behind him was a wall, and before him lurched a threatening figure, a voice from the darkness asking inadmissible, impossible questions. His hand unconsciously touched the watch in his vest pocket. This was not some eccentric or madman. Oh, no. Something else was going on here, something unreal, like a dream. Maybe I'm dreaming, he thought with hope. He did not wake up, however, as always happens when this question is asked in a dream.

"What would be the use even if I did want to, sir? It can't be done. I mean, all right, maybe time isn't, as you described, sir, a river, I don't contest that, but that...tree...with the forks in the branches...and the rest. The double... But how can a person ever get the chance to change anything? Go back to the past?"

There was no reply from the shadow. The seconds lapsed, long, silent, full of expectation. And then, instead of the stranger, the wall quartet suddenly resounded, breaking off the tense silence and prompting the old man for the first time in his life to jump at the harmonious announcement of the full hour; the very next moment it was transformed into a discordant confusion of grumbling, chirping, chiming, and waltz music.

The visitor remained motionless until the last echo of the grenadier's bass died out and then with a rapid movement placed the top of his cane next to the pocket watch that lay on the illuminated felt counter.

"You will look at it, won't you?"

A deep sigh of relief escaped from the old man, as though a heavy load had been taken off his chest. His eager hands finally caught hold of the precious object; they started to turn it over and feel it, examining it as carefully as eyes could.

"Certainly, certainly. Rest assured, sir. Right away. It's not too late. If you would be so kind as to come by in the morning. As soon as I open. It will be ready. At your service, sir. At your service."

The foreigner abruptly turned on his heel, missing the watchmaker's humble bow. The sound of the cloak's stiff fabric merged with the ringing of the bells and the closing of the door. The tall shadow passed quickly in front of the store window and disappeared down the street.

The old man slowly sat down on the chair next to his workbench and put the pocket watch upon the rubber surface. He gazed at it for a few moments, turning it over curiously, and then reached to open the lid.

But he did not complete the movement, for he suddenly realized that in the excitement of the moment before, he had forgotten something: he had not given the customer a receipt for the watch. Inexcusable, he thought. That had never happened to him before. All right, he had been disoriented by the visitor's unusual appearance, by that strange story, but even so! An unpardonable oversight for a watchmaker who cares about his reputation. What would the foreigner think of him?

He grabbed the receipt book and a pen from the counter and rushed toward the door with stiff movements. Suddenly disturbed, the bells above the door protested sharply. Outside it was cool and windy, a November evening at the foot of a mountain which already carried a great cap of snow. Shivering for a moment, the watchmaker looked for the visitor down the row of streetlights. But no one was there. Perplexed, he turned and looked in the other direction. Just as empty.

He stayed in front of the shop a little longer, turning back and forth in disbelief, and then returned inside. Where had he gone? Had a carriage been waiting for him nearby? But no, he hadn't heard anything. Standing at the door a moment, the old man finally shrugged his shoulders. He would apologize to the foreigner for this oversight when he came in the morning. In any case, it would make no difference then. The most important thing was for him to take care of the watch.

He returned to his workbench, interlaced his fingers and cracked his knuckles like a pianist before a performance, and then drew up the squeaky stool. Before he pressed the clasp to open the lid, he briefly rubbed his fingertips with his thumbs.

His eyes first went to the inner side of the lid. It was an inadvertent, almost automatic act: that was what he always did with the other pocket watch that he kept with him always. There was an engraved inscription there as well that for some reason seemed familiar to him. TO J. FROM Z. was on the gold-plated concave circle, and several long moments had to pass before the old man realized what it was. The shape of the letters, of course! The same large, ornate letters as... But how was it possible?

And then there was no more time for ordinary amazement; on top of the wreath woven of twelve elongated Roman numerals the two black hands had started their crazy dance.

II

They seemed to have a will of their own, moving by themselves—but in the wrong direction. They started to turn backward, as though measuring the past, first slowly, so he could follow them, then faster and faster. The watchmaker instinctively withdrew his hands from the activated watch, but his eyes stayed riveted to its face.

He stared at the big hand as it accelerated and then finally disappeared, transforming into an excited circle; it looked like some sort of film had been placed over the face. The spinning of the small hand was perceptible somewhat longer, and then it, too, melted into an indistinct veil.

This tremendous spinning made the watch tremble on the rubber surface. It suddenly occurred to the old man that he could stop the magic if he closed the cover, but he did not have the courage to touch it. Holding tightly to the edge of the workbench, he felt that the accelerating vibrations of the watch were being transferred to his body: he, too, was shaking as though he had a fever.

And then the trembling stopped, for the watch had detached from the tabletop and started to float slightly above it. Although it was illuminated by the strong lamplight, no shadow lay beneath it, just as though it were transparent. A high, shrill whistle started to sound, almost at the upper threshold of audibility; there was something unsettling in that sound, and the old man wanted to put his hands over his ears but was unable to do so.

As though bewitched, he simply stared at the floating object before him that continued to rise slowly until it reached the height of the old man's eyes. It rested there a few moments, hesitating as though thinking what to do, and then started to spin around its vertical axis. Just as with the hands on the face, the spinning became faster and faster until there soon formed the illusion of a small ball before the watchmaker's bewildered, slack-jawed face.

As though cut with countless facets, the ball first brightly reflected the light from the lamp on the workbench and then began to radiate its own light as the shrill noise became louder and louder. To the old man's relief, the unbearable sound soon rose above the frequency audible to human ears, leaving behind a muffled, almost palpable silence.

In just a few moments, the dull grayness turned into a reddish glow, then into yellow heat, and finally there was a rapid sequence of shades of white, rushing to the inevitable climax, the act of release. The old man greeted this orgasm of light with wide open eyes, unable to lower his eyelids; in any case, what could thin, wrinkled skin do against the uncontained fierceness of a summer sun less than a foot from his head?

Although he was completely blinded by the explosive flash, he did not feel any pain or even discomfort. The only thing he felt was the strange sensation of being in the middle of an endless emptiness, impenetrable and silent; he made his way through it effortlessly since there was no base or support to hold him back. His body seemed to have lost all weight and along with it all sense of direction: up might be down, or somewhere to the side—he was not able to distinguish anything.

Is this death? he wondered. If it is, then it is very mild, even pleasant. Like a dream. This was not how he had imagined it. Actually, he had not imagined it at all. Who imagines what death looks like, anyway? He had the vague feeling that he should be afraid for some reason, but instead of fear or at least discomfiture, he was filled with childish curiosity. Where was he? Would he remain incorporeal like this forever? Did time exist here? Why couldn't he see or hear anything?

As if in answer to this last question, sounds started to come from a great distance. He did not recognize them at first; they were too muffled. At first they resembled the scraping sound of gravel being rolled by waves on the shore and then the drumming of rain on the leaves in a forest on a wet spring evening. Then something in their rhythm seemed not only recognizable but familiar: the monotonous, regular repetition, harmonious only in the introductory chord and then completely dissonant...

There were seven strokes from the moment he started instinctively to count the hour sounding in four disparate registers. How many had he missed until he understood what it was? Three—or maybe more? There was only one way to find out, although he did not understand why it was important to ascertain this fact. He reached for his vest pocket, forgetting completely that he had become incorporeal. But the pocket was there, real and tangible, as were his vest and hand—everything was there except the watch that Mary had given him, that day... The watch had gone!

How was that possible? Why, a little while ago... He looked at his vest in panic, only realizing when he saw it that his sight had returned. He was no longer blinded or surrounded by impenetrable emptiness. He stared at his body for a few moments, filled with disbelief, and then slowly raised his eyes and looked around himself.

He did not notice what was wrong right away. Everything seemed to be normal: things were in their proper places—the workbench which he was still grasping convulsively with one hand, the counter covered with green felt, the old-fashioned clothes tree in the corner with his winter coat hanging from it, two armchairs with reddish upholstery and the round coffee table between them with its thin, curved legs, the large grandfather clock with its pendulum, the mirror on the opposite wall with the black wrought iron frame...

Only after he had taken all of it in did he realize where the problem lay: he should not be able to see it all. The only light in the shop was from the small lamp with the green shade in front of him, and the lamplight barely reached the counter. Now, however, he could see everything as clear as daylight...

Day!

Daylight flooded through the large shop window with WATCHMAKER written in an arch of dark-blue letters. It was bright and clear, light that in this region was seen only in late spring and during the short summer, certainly not in mid-November. But it was not early winter outside; when a little girl skipped past the shop soon afterward, the watchmaker was perplexed to see that she was wearing a checkered dress with short, ruffled sleeves.

He got up from his workbench, finally lifting his numb hand from its edge, and took slow, hesitating steps from the counter toward the entrance. When he was in the middle of the shop, out of the corner of his eye he noticed something moving to his right and turned slowly in that direction, encountering his own reflection in the long mirror.

He squinted and stared at his image, refusing to believe what he saw. It was he, without a doubt, but different, changed—rejuvenated. The person returning his look from the glass was not an old man, stooped, his forehead full of wrinkles, gray-haired and balding. He was a young man, barely thirty, standing straight, with smooth skin and thick, dark hair.

He started to touch his face gingerly, afraid that even the lightest pressure might deform it into its former deteriorated grimace like a wax mask. His fingers slid over his mouth, chin, cheeks, striving to feel the trickery, but there was no deception: his youthfulness was real—as real as everything else around him seemed to be.

He continued to look at the long-forgotten person in the mirror, while the confusion in him slowly withdrew before the mounting excitement, when suddenly, like a bolt of lightning, he had a sensation that he had experienced only a few times before, but never as strongly. The feeling of *déjà vu* was all-encompassing, overwhelming: he had stood on this same spot before, looking at himself in the mirror, and the bright summer day had been exactly the same.

Something caught in his throat when he realized what had to happen next. He had no doubt that it would actually happen as he quickly turned around to face the entrance. The bells above it started to fly loudly in all directions that same instant. Only she entered like that: like a whirlwind of blond curls, with her long, rustling dress, her smile so enchanting in its radiant cheerfulness...

Mary!

He knew that she would not turn to look at his wide open eyes, would not notice the paralysis that had come over him, that she would not hear the thunderous drumming of his heart that so filled his ears he felt as though the whole shop were echoing. He knew that she would rush to one of the armchairs and unload the armful of colorful boxes she was carrying.

Her words reverberated in his head a moment before she uttered them, like a reversed echo that precedes the original sound.

"It's so terribly hot. It's even worse downtown. And crowded. You have no idea. It's as if the whole town were outdoors. You should have come with me. You sit inside too much. It's not good for you. You could have closed the shop today. There are a lot of people here, too. You should see how many carriages there are in the square. Goodness, I'm all sweaty. And I'm terribly thirsty."

She started to rummage impatiently through her rather large handbag made of flowery waterproof fabric; it was always full and now seemed truly inflated. A full minute went by before she finally found what she was looking for. The small box was wrapped in shiny green paper, and the turquoise ribbon had curled ends.

He did not have to open it to find out what was inside. Nevertheless, he did it as inquisitively as he had earlier because he was impelled by the inexorable pressure of *déjà vu*. Once he had lifted the cover of the pocket watch and looked at the engraved inscription, he smiled broadly and said the sentence he knew went at that place.

"It's beautiful. Thank you."

He did not have the courage to be more eloquent in his thanks this time, even though he wanted to with all his heart. The object he held in his hand meant more to him than a present from his fiancée: it was an infinitely precious keepsake from which he had never parted in the many years which followed. Even so, the fear prevailed that if he used any other words he would cause an irreparable disturbance and would lose this feeling of *déjà vu* that was guiding him.

Mary returned his smile and then went up to him, raised herself on tiptoes, and kissed him. It was a light, brief touch of the lips, on the very edge of decorum, considering the time and place, but it made him tremble nonetheless. She suddenly turned toward the door, feeling awkward, to see if anyone was about to enter, and then began to pick up the boxes from the armchair. They were full of the beautiful things she had chosen to look stunning at the upcoming ceremony.

"I'm going to take all of this and get changed. I'm all sweaty and sticky. It's so hot. You should put on something lighter, too. You'll boil. Let's go have lunch at the Golden Jug. What do you say? It's the coolest there right now, in the garden under the linden trees. All this shopping has made me hungry."

She smiled at him again, a special mixture of affection and apology, and then rustled in her whirlwind manner toward the door—to meet the inevitable. The sequence of events stood before him, completely clear, illuminated by the powerful beacon of *déjà vu*: the wild music of the horse bells briefly muffling the thudding that was rapidly approaching; her hurried departure onto the pavement in front of the shop as the empty carriage jumped wildly on the cobblestones; incautiously crossing the street at the very moment the confused horses without a driver, left too long in the sun and frightened by who knew what, could no longer be stopped; the horrible shock at realizing that there was no way of escape; someone's scream from the other side of the street that seemed to last an eternity; and then the multicolored boxes flying in all directions, opening up and spilling their insides: an elegant lemon-colored dress with an abundance of lace, a yellow hat with a large brim and a wide ribbon tied in a bow, shoes with large, shiny buckles, a pile of silk undergarments that certainly should not have been displayed like this—the senseless nakedness of death.

"Mary!"

He had to overcome the violent river to utter this word, to scrape off the previous deposit on the palimpsest with his nails, to seize hammer and chisel to write a new inscription on the virgin surface of the granite. The magic of *déjà vu* shattered at that moment—there was no room for this call; his role had been to remain silent, to follow her out merely with his eyes. Stepping out of the play in which he was unwillingly acting, he was suddenly alone, exposed to the winds of time without a guide to light his way, but also without the ominous inexorability of the predetermined.

She stopped at the door and turned. "Yes, Joseph?"

He didn't know what to say. He certainly could not start explaining, particularly since he himself barely understood. So he simply went up to her and hugged her, together with her armful of boxes. It was a hard, awkward squeeze, calculated above all to keep her there, not to let her leave. He knew that this could arouse her suspicions, since such public displays of intimacy were not at all characteristic of him, but he chose the lesser of the two consequences.

"Oh, Joseph, dear, someone might come," she said in a voice whose reproach was only feigned. "Be patient a little longer..."

Somewhere at the top of the street, from the direction of the square, dull thudding could be heard. It approached rapidly, mixing with the clatter of bouncing wheels. The sound was similar to thunder heard backward—from the dying out to the explosion. Mary tried to wriggle out and turn toward the window, but Joseph's embrace held her tightly.

"What was that?" she asked, turning her head to the side.

"Nothing...a carriage, probably...in a hurry..."

If there was an end to his sentence, it was lost in the deafening stampede, in the strike of lightning. Just like the shadow of a low cloud, the unbridled team whizzed past the watchmaker's shop in a whirlwind of hooves, wheels, manes, empty driver's seat, foaming muzzles, spinning axles, terrified eyes, reins dragging on the ground, sweaty crupper—and afterwards the thunder resumed its natural course again.

"Someone could get run over," said Mary, after Joseph's squeeze finally relaxed. Now he was standing almost penitently next to her, not knowing what to do with his hands that had held her like a vise a moment before.

"Carriage drivers have become so inconsiderate, even arrogant. You should see them down in the town. They tear around like madmen. And how they whip those poor animals. It's terrible."

"No one will get run over, Mary. Not anymore."

She looked at him suspiciously, confused by the changed tone of his voice. He had said it too seriously, as though pronouncing some kind of oath. Even so, as he uttered them, he was aware that they were merely empty words of comfort similar to those said to calm a child the first time he asks about death.

Of course, someone would get run over. The inscription chiseled in granite could not be erased. On another fork of the tree of time he was now running into the street and bending in a convulsion of pain over the unmoving body, while tufts of yellow fluttered all around. He could pretend that this no longer concerned him, that he was now safe on this branch where Mary was standing next to him, the very incarnation of the vibrancy of life, sweaty, laughing, thirsty. But although he did not understand it much, the realization that both courses were equally real was painfully clear to him.

The clarity with which he remembered the anguish he felt as he lifted her off the bloody pavement, heavy with lifelessness, the hopeless insensibility into which he then plunged for a long time afterward, the slow succession of months and years filled with the deceptive oblivion brought by tedious work, and the lonely, nightmare-filled nights in which the past relentlessly visited him, until that far-off November evening when the bells suddenly rang above the dark door to announce the arrival of the mysterious visitor—that clarity, that hard certainty of memory was the price he had to pay for this unique privilege that he had been given for who knew what reason: to return to a past time and undo the effects of cruel chance.

He knew that this price did not give him the right to be dissatisfied. On the contrary, the shadow over his restored happiness was a very thin, transparent veil. Nevertheless, in the years that followed, only Mary's intoxicating, infectious cheerfulness managed to dispel the mask of melancholy that periodically and for no apparent reason covered Joseph's face.

The Artist

I

He unlocked the door and entered the room.

If it were not for the bars on the window, it would have looked just like an artist's studio. The half-open window with the thick drapes and pleated curtains rose almost to the ceiling, letting in an abundance of light during the day. Painted white, the bars were not too conspicuous, but they could not be overlooked. They were not there to prevent anyone from escaping, for this was not a prison, but rather to prevent the final retreat that the mind of the room's occupant might seek from its own darkness.

The room was sparsely furnished. To the right of the window, at a slant, stood a rather large easel spotted with dried streaks of paint and placed on a covering of newspapers, yellowed from long exposure to the sun. Next to the wooden easel was a tall, thin chair with a low back and rungs for feet. Part of the lower half of the wall nearby was covered with mounted shelves that held a disarray of art supplies: mostly squeezed-out tubes of paint, half-empty little bottles of paint thinner, brushes of different sizes, dirty palettes, a bunch of used charcoal sticks and pencils, soiled flannel rags, large sketch pads, a pile of rolled-up canvases, and several cans with bright labels and no lids.

The only light source turned on in the room was a reflector light on a short support attached to the middle of the ceiling. The narrow beam illuminated the canvas on the easel, reflecting brightly off the fresh layer of paint. The edges of the beam that reached the uncovered floor glistened off the polished parquet.

He headed toward the other side of the room and sat on the end of a narrow bed with a brass frame, next to the door that led to the small bathroom. In addition to the bed, there was only a little white table with drawers: on it was a lamp with a yellow canvas shade, a vase with large-petaled purple flowers, and an old book with a black cover, pink-edged pages, and a wide ribbon as a bookmark.

His eyes went to the wall facing the window. He could not see well in the semidarkness, but it made no difference. He knew what was there: three paintings in simple gray frames, unevenly arranged. Three scenes of darkness disrupted in the middle by a beam of light: the flickering glow of a torch in the corridor in front of a cell, cone-shaped lamplight illuminating a jumble of old things on an office desk, the green glow of the felt on a watchmaker's counter. And outside the beam, distinct from the surrounding shadows like a concentration of the night, was a spectral figure without a face.

"Good evening, Doctor." She said it softly, with her back turned, sitting on the tall chair. All she had on was a short-sleeved nightgown; her fragile shape could be discerned through its thin, semitransparent fabric. The scene was not stable because the light material trembled and fluttered under the gusts of warm breeze from the window. Her bare feet with their small toes were resting on one of the rungs. The brush in her left hand was making rapid, short strokes about the canvas.

"Good evening, Magdalena. The nurse tells me that you are painting again?"

"Yes."

"Isn't it a bit late for that? Wouldn't it be better for you to go to bed and then get down to work tomorrow morning?"

"I can't. I have to finish the painting as soon as possible."

"You were never in a hurry before."

"Now I have to."

"What for?"

"He was here."

The doctor closed his eyes a moment and drew his fingertips across his forehead. "He came to visit you again?"

"Yes."

"Did he tell you a new story?"

"Yes. The last."

"The last?"

"There will not be any more."

"Oh? Why?"

She did not answer right away. In the silence that descended, distant sounds of the summer night were suddenly audible: the soft rustle of leaves in the tops of the tall trees surrounding the sanatorium, the idle chatter of crickets in the grass, the sharp call of a bird.

"He's leaving."

"Is that why you are in a hurry?"

"Yes. I want him to see how I have painted him. He promised he would come one more time just for that."

"You are going to paint him? He finally showed himself to you?"

"Yes."

"But he has always remained hidden before. You never once saw him during an earlier visit. That is why he has no face in your paintings. Why the change now?"

"He will still remain hidden."

"How can that be if you paint him?"

Before she replied, she dipped her brush in the paint on her palette, mixing colors for several long moments.

"I'll paint him, yes," she said at last, returning the brush to the canvas. "I'll even tell you all about him, if you wish. But of course, you won't believe me."

"Why do you think that?"

"Because you think I'm crazy." She said it evenly, as though stating the obvious. "My madness conceals him. Better than any darkness."

"You know that we do not use such words here."

"I know. You have other, milder expressions. But that does not change the essence of the matter. There are still bars on my window, and you keep the door locked."

"The bars are there for your own good."

"So I don't lean out too far by accident and fall?"

"Accidents do happen."

She put her head close to the canvas for a moment, engrossed in painting some detail. "So, then, you could believe me."

"I could listen to you and then judge."

"That's fair." She moved back from the easel, taking a look at the detail. "Tell me, what do you think—who is he?"

"How would I know that?"

"But you certainly have some idea," she said, searching again for the proper color on her palette. "I have told you about our meetings. You know his stories."

"Someone very powerful, obviously, since he can do whatever he wants with time."

She found the right color, and her bared left arm started to move quickly before the canvas once again. "The devil?"

For several moments he silently watched her fluttering figure before the painting she was working on.

"He would be a very unusual devil," he said at last. "A devil who does good deeds without any recompense."

"Do you think he did the right thing?"

"Didn't he? Three unhappy people received a unique time gift, as far as I understood."

"And now they are less unhappy?"

"Why, I suppose. They should be. Particularly since they were not asked for anything in return."

"He, too, thought he would make them happy. At first."

"He doesn't think so anymore?"

"No. That is why he is leaving. He discovered that it is truly the work of the devil to fool around with time, even when you have the best of intentions."

"Where did he go wrong?"

She put her palette and brush under the easel, threw back her head, and tried to shake back her long hair. But the curly, auburn locks were too tangled from long lack of combing.

"Do you remember the story about the astronomer?" Without turning around she pointed her thumb to the right, to one of the three paintings on the wall. "If it hadn't been for his nighttime visit before the execution, Lazar would have happily gone to the stake, convinced of how correct, even exalted his sacrifice would be."

"But it was a mistake. Visiting the future showed him that his sacrifice had no meaning."

"Do you think that people should be freed from their mistakes? Even when it ends up destroying their happiness?"

"Happiness based on illusion, deception?"

"And what happiness isn't?"

He did not know how to reply at first. He felt like a chess player whose opponent has made what seems like a quiet move, but one riddled with hidden traps.

"What is the meaning of happiness if it entails the loss of a life?" he asked at last, in a muffled voice.

"And what is the meaning of life without happiness? That is the impossible choice Lazar was forced to make. With the best of intentions. Everything would have been much simpler if he had not seen the future."

"Your visitor did not tell the story to the end. He did not tell you what the astronomer chose."

"He didn't because it made no difference." She stopped a moment. "What would you have chosen if you were in his place?"

A somewhat stronger gust of air from the window raised the hem of the nightgown, revealing slender calves. It brought into the room the abundant smell of grass and certain traces of ozone—the first sign of the storm that was on the way.

"And what about the professor of paleolinguistics?" he asked, avoiding any reply. He raised his eyes inadvertently to the second painting on the shadowy wall. "She has no cause for regret because chance was thwarted; on the contrary, she went back to paradise."

The artist did not reply at once. She leaned toward the shelf behind the easel, started rummaging around in the tubes, selected one and squeezed out a bit of the contents onto her palette. Then she took the flannel rag and wiped off the tips of her fingers.

"To a paradise she was denied, actually. Eva was only an observer in paradise, without the chance to take part in it."

"I didn't have the impression that she felt it was unpleasant being...a ghost. Many of those studying the past would be ready to give half their lives, even more, just to be in her position."

She began applying more paint to the canvas. Now she was working on the middle of the painting. "She would have given it all up just for one sip of heavenly tea."

"Perhaps, but that was the price she had to pay. There was no other way to find out if everything she had written was accurate."

"But imagine if it turned out that she was wrong. That primeval language was quite different from what she thought. It would be a twofold defeat: she would have squandered her past life, and before her would be a paradise that she could not enter."

"It didn't have to be that way. She might have been proved right."

"Would that be enough comfort for unattainable paradise?"

"But if she didn't return to the past, she would have been left in doubt until the end of her life. That way at least she found out where she stood."

"Isn't it actually uncertainty that makes life possible?" Another quiet move full of hidden menace.

"Your visitor didn't tell you the end of that story, either," he said after a slight hesitation.

"For the same reason as before. It makes no difference what Eva hears when she gets to the fire. The best thing for her would be never to have left her basement office."

A blue flash suddenly appeared in the upper part of the window, but no thunder was heard. The storm was still some way off. Only the choir of crickets seemed to accelerate its chattering tune.

"The third story differs from the first two in this regard," he said, again turning to the paintings on the wall. "There is no uncertainty in the end."

"No, there isn't, but it still is not a happy ending, as it should be."

"It isn't?"

She turned her head toward the window and stared at the darkness.

"It's really sultry," she said. "I can hardly wait for the rain. It's hard to paint in this heat. I'm all sweaty."

He closed his eyes again and started to make little circles on his temples with his fingers. That was where he first felt the change in weather. The dull throbbing there that was slowly spreading to the back of his head indicated that he would spend the night wrestling with a headache.

"It isn't," she continued. "Perhaps it would be if he did not have the memory of the other stream of time in which Mary died."

"But, actually, that was not the memory of something real. It was more the recollection of a bad dream."

"It lasted too long to be just a dream. More than a quarter of a century. Why was it necessary to let Joseph suffer so long? If it was possible to help him, and someone was willing, he should have been put on the other branch of time right after the accident. Only then could it all have looked like a bad dream. This way the scars were too deep and real."

"Why wasn't that done? Did you ask your visitor?"

"Yes, of course."

"And? What did he reply?"

She drew the back of her hand across her forehead. "He said he could not have done otherwise because then the story would not be as good. If he had offered his time gift earlier, the hero certainly would have had a better time of it, but then the story would be weaker. The same holds for the other two."

"Strange. I had no idea that the devil cared so much for literary effect."

She stopped with her brush in midair, not finishing the stroke. "He's no devil, of course. If he were the devil, he wouldn't care at all about what happens to his heroes. And he is abandoning his time stories just so he doesn't transgress against them anymore."

"Well, who is he, then, if he isn't the devil?"

A dull roar finally broke the tranquility of the summer evening. The storm was about to break. As if by some inaudible command, the crickets suddenly fell silent.

"Wasn't it clear from the very beginning? The one who tells stories. The storyteller. The writer."

"The writer?" he repeated obtusely.

"The writer, yes. The writer who accepts responsibility without which his divine omnipotence becomes just unrestrained diabolic self-will."

"Responsibility toward whom? The heroes of his stories? But they don't exist, they are not real people. There is no reason to burden your conscience because of them."

The thin, pleated curtain on the tall window suddenly billowed out like a white sail. Leaves rustled sharply in the nearby treetops, and the edges of the old newspapers under the easel started to flutter restlessly.

"Do you think so?" she asked briefly, turning her face toward the fresh air entering from outside.

He pinched the bridge of his nose firmly with his thumb and forefinger. The pain from his temples had moved there, becoming more piercing, burning. This conversation had to be terminated. They had reached a dead end, and it was already quite late. They would continue the next day, when he was rested.

"So the writer is leaving us," he said, getting up slowly from the end of the bed. "There will be no more of his mysterious visits." He headed for the door, and then stopped, remembering something. "By the way, did he tell you how he managed to enter your room and then leave it, in spite of the bars on the window and the locked door? Did he perhaps transfer the omnipotence he has in his stories to reality?"

As soon as he said this, he thought that the question had not been formulated very skillfully. The fatigue and headache were clearly having an effect. She might think he was making fun of her, which would not be at all good for their relationship. It had taken him a long time to get her to leave the cocoon of silence in which she had enclosed herself, to start telling him about the pictures she painted.

"The storyteller cannot transfer his omnipotence to reality," she replied. There was no trace of rancor in her voice. On the contrary, it had a note of joy in it, probably from the excitement of the approaching storm. That often happened among the patients. It was as if they were permeated with the electricity that filled the air. The nurses would have their hands full tonight.

"Then how?"

Outside, it started to rain. The drops were still scattered, but their heavy drumming indicated that they were large, stormy.

"Don't you get it?" she asked. "There is only one other possibility."

He stared fixedly at her back, over which the thin nightgown was now wrinkling like ripples on the surface of the water. "I don't get it. What possibility?"

"This is not reality. This is also one of his stories."

He stood immobile in the middle of the room. He knew he should say something, that it was expected of him, but he could by no means find the proper words. He was confused not by what she said but by how she said it. That flat voice again, as though it were stating the obvious.

"I said you wouldn't believe me."

He snapped out of his paralysis. "It's not easy to believe. Would you believe it if you were in my place?"

"Oh, I would, certainly. It's not hard for me. I'm crazy, right? But you aren't. In addition, you are a man of doubt and not belief. You'll still be suspicious even after you see the proof."

"Proof?"

She laid the palette and brush under the easel again, wiped her hands on the spotted flannel rag, and reached for something in her lap. A moment later she turned to face him in the tall chair and raised her hand in the air. A yellow gleam danced in the bright reflector beam.

"Where did you get that?" he asked, squinting at the pocket watch.

"From the writer, of course. It is his gift. There is a dedication engraved on the back. Here, take a look."

She held out her hand with the watch, but he did not take it right away. He stared at the golden object on her palm, feeling the hairs bristle on the back of his neck. Everything is really full of static electricity, he thought. As if in reply, everything in the room suddenly flashed a blinding blue. He knew what would follow, but the violent explosion that resounded just a fraction of a second later still made him start.

She did not even blink, as though completely deaf.

"It's only thunder, don't be afraid," she said gently. "Go ahead and take the watch."

He did it hesitantly, timidly. It was heavier than he expected, convex on the top and flat on the bottom. His fingers felt the engraving on the back, and he turned it over in his hand. The inscription was tiny and curling, calligraphic. Two names above the middle of the circle. Hers, his.

"So that is the name of the writer," he said. It was something between a statement and a question.

She did not reply. The silence that reigned was disturbed only by the downpour from the low clouds. Periodic flashes of lightning illuminated the curtain of water just outside the bars. The rain was falling straight down, so the parquet before the window was completely dry. The air in the room was saturated with humidity and some new, pungent smells.

He started to turn the watch in his hand, looking for the latch that would open the lid.

"Are you sure you want to open it?" she asked quietly.

He found the latch but did not touch it. "Shouldn't I?"

"No, if you're not ready for a time gift."

"What kind of time gift could I receive?"

"One with the power to alter your entire life."

He smiled. "Is there something like that for me? There is no execution awaiting me at dawn, neither am I plagued by doubt in my old age as to whether everything I have done has been mistaken, and I don't have the least dark spot in my past that should be removed."

"Oh, it exists for everyone. Even someone crazy like me. It is the final time gift."

"Final?"

"That's right. Tell me, what is the only thing you know for certain about your future?"

He thought for a moment, looking at her suspiciously. "That I will die, if that is what you are thinking of."

"Yes. But you don't know when it will be, tomorrow or many years from now, right? And it is this very ignorance that allows you to suppress the awareness of your own mortality which would otherwise become an unbearable burden. Not knowing when you will die—that is life's main stronghold."

The third quiet move. He had the strange impression that an invisible net of checkmate was woven around him and there was no escape. "And if I raised the lid I would find out?"

"You would find out."

The jets of rain were suddenly slanted by the wind. They started to soak the curtains and drapes and made puddles on the floor under the window, reaching all the way to the newspaper under the easel.

He glanced in that direction and then turned toward her again.

"How do you know?"

"Because I opened the watch."

"And you found out when you will die?"

"Yes."

He shook his head slowly. "There's something amiss there. Didn't you say that your...writer...stopped giving time gifts after the bad experiences he had with them in the first three stories? That he had accepted the responsibility that goes along with omnipotence, that his conscience did not allow him to inflict harm on his protagonists? If this is his story, too, as you claim, then he has behaved very badly toward you: you have received the cruelest of all time gifts."

"I received it because I asked for it myself. He fought against giving it to me for a long time."

"But why did you ask for it? Didn't you just say there is no sense in finding out when you will die?"

"My case is special. If he had already imagined me crazy, then I had the right to know how much longer I will have to be like that. It was the least he could do for me, although he did not find it easy."

"What about me?" he asked, rubbing his forehead with his fingers. "I can raise the lid, too. What would be his justification for me if I, too, am a character in his story?"

"You can, yes, but even so you won't do it."

"Why not? What's stopping me?"

"The writer's omnipotence, of course. He won't let you do it."

A smile spread over her face. His thumb started slowly toward the latch of the pocket watch, but the movement was not completed. The knock that suddenly resounded was short, sharp. The tall nurse did not wait to be given permission to enter. She stopped by the door and said quickly, "Ah, you're still here, Doctor. They need you urgently in room forty-three."

He stood there without moving, holding the watch in his open palm. A gust of wind filled the wet curtains again and lifted the newspaper all the way to the easel legs.

"Why, everything is open here," muttered the nurse, rushing toward the window. "Magdalena, you'll catch cold; put something around your shoulders."

Continuing to smile, the artist slowly took the watch from the doctor's hand.

"Hurry, they're waiting for you," she said gently. "We'll see each other tomorrow. There is plenty of time. The story ends here for you, but we will talk about it for some time to come."

II

The nocturnal storm had long since passed, leaving behind dense humidity full of the smell of decay that would float over the wet soil until the sun rose in a short while. Flattened blades of grass started to straighten up slowly, throwing off the remaining drops of water, but here and there occasional dripping from the leaves bent them to the ground again.

The wind had stopped altogether, so that the deep silence just before dawn was disturbed only by short birdcalls. They sounded inquisitive and fearful, like the calls of lost shipwreck victims on the high seas. The spectral echo of these cries remained in the motionless air long after the original sound had died out.

In the vague light of dawn that filled the large window, the white bars no longer stood out against the curtain of darkness. The milky morning light also dulled the sharpness of the reflector beam illuminating the canvas, making it milder, paler. The contours of the few objects in the room seemed to lose their solidity in this new light.

She was still sitting in the tall chair in front of the easel, staring at the canvas before her. She had put a terry cloth bathrobe the color of a ripe lemon over her nightgown, which made her look even smaller because it was too big: the hem reached almost to the floor, hiding the rungs and her bare feet, and the sleeves hid her hands completely. The cuffs were stained with paint and seemed to merge with the palette and the brush held by her invisible fingers.

The enormous face of the pocket watch covered the entire canvas, almost reaching the edge at four points. The corner sections outside this surface were mere dark voids that would certainly have been left out if the frame had been circular. Although the thickness of the large watch could have been neglected as well, since it was not part of the area encompassed by the circle, it was still indicated: a barely noticeable reflection of light from some unseen source conjured the gentle curve on the edge.

Compared to the surrounding dark tones, the bright central whiteness almost burned the eyes with its cold glow, sharply emphasizing each detail on it. The twelve numerals were long and thin but not regular. They looked unstable, as though a restless flow of water were passing over them, making them bend and twist. The rippling was more distinct in some places, bending parts of the numbers into senseless shapes or pushing them all the way over the edge of the face.

The four hands were of the same length. The pointed ends reached the perimeter, widening toward the center just like narrow, elongated fern leaves, with a small slit in the middle. The leaves ended in thin stems that met in one point, as though sprouting from the same bud. The opposite hands formed two segments at right angles to each other. The vertical pair linked the numbers twelve and six, the horizontal nine and three.

A semitransparent body was resting on these crossed hands, following their shape. Its arms were stretched over the horizontal hands, tightly attached to them. On the palms of each white glove bloomed a large red stain although there were no nails. The fingers were clenched like claws but did not reach the red blossom.

Red spots spread also on the dark leather shoes, but they looked less conspicuous there. The pain inflicted by the unseen nails was manifested in the unnatural angle of the legs, whose pierced feet were trying in vain to lighten the load of a body without support.

The long cloak was covered at the bottom by a layer of dried mud, depicting clashing brown smudges on the black background. The edges of the cloak were worn and shabby, the hem unsewn in places. The lining was a fiery color and was torn in one place as though it had gotten caught on a thorn bush.

In front of the torso was a dark cane with the top downwards. It floated vertically without any support, casting a slight shadow over the surrounding whiteness. The ivory hourglass at its end was cracked in the middle. It seemed as though all the golden sand had poured out of this crack, leaving just an empty shell that could no longer measure anything.

The tall hat was covered by a film of fine dust, subduing the black silky shine. The shape of the derby was ruined by several uneven dents. The wide brim no longer concealed the face because the light came from below, yet it still could not be seen. The emptiness of virgin canvas gaped in the place where it should have been.

She knew she had to finish the painting, that the time of the last story was running out. The missing face was there before her eyes, perfectly clear in its repentant agony, but the fingers in the sleeve refused to lift the brush.

She had imagined the scene quite differently. She had wanted to paint him doing what he always did during his earlier visits. He would appear soundlessly at the bathroom door, but she would sense his arrival even though she was sitting at the easel with her back turned. He would take the hairbrush with the broad handle of lacquered walnut from the shelf under the mirror in the bathroom. It had stiff sharp bristles, which was what her tangled hair required.

He would brush her hair patiently and at length, just as long as it took to tell a story. When he reached the end, her hair would be loose and smooth, and the disorderly curls would be turned into a graceful row of waves. After the very first brushing, she no longer allowed anyone else to brush her hair and did not do it herself, either. She would wash it regularly but would leave it uncombed between stories. The nurses did not try too hard to dissuade her, seeing in her stubborn insistence just one of the caprices of their special patients.

It would be a nice painting, perhaps the prettiest of all four. But this still could not be a love story. Or not just that. It came at the end, after the others, linking them into a whole, so that it had to talk about redemption much more than about love. She had realized that necessity but could not understand why redemption had to be ultimately so painful. As she was painting, she herself had felt the torment of the rusty nails piercing the tender tissue of her hands and feet. She had somehow managed to endure the nailing while it was impersonal. Now, however, the crucified person had finally to receive a face.

When she started to make short, rapid strokes on the only unpainted part of the canvas, her eyes glazed over and her lips drew together with a slight tremble. But her hand was sure. From the seemingly unconnected lines, the oval emptiness started to take the shape of the writer's face, distorted by the primordial sin of his art.

And at that moment she understood why the pain was necessary. Without it, he would only be an indifferent god who justified the harm he did with good intentions. If he justified it at all. The suffering he chose brought him redemption by making him identical to those he had transgressed against. Without this sacrifice it would not be possible to accept the final responsibility that goes with writing.

When she had painted the last stroke, she slowly leaned her head backward, and her long, auburn hair spilled down her back. As before, it was a movement of ultimate intimacy, of surrender. She closed her eyes in anticipation. Somewhere outside echoed a protracted, joyous chirp, and the paleness of dawn was edged in pink.

The brush sank into the hair on the crown of her head. The curly locks were too tangled, so the combing out inflicted pain at first, although her radiance disavowed it. The walnut-handled brush made its way slowly, with short strokes, going back a bit whenever the tangle of wild waves offered greater resistance. The lower it got through the agitated sea, the harder and slower was the progress, and at the very bottom the curls were almost matted.

When her hair was finally untangled, the arc of the sun had already pierced the porous green of the treetops. The brush was raised again and this time sank smoothly into luxuriant waves. It made its way easily, straightening out the last rough spots, taming the most obstinate curls. Even though the ends were no longer matted, it stopped there a moment, unwilling to leave the locks that now seemed to have absorbed it. But this moment of hesitation quickly passed. When it slipped out, the curled ends rebounded as though on hidden springs.

She remained immobile, her head thrown back. The slanted morning rays pierced her closed eyelids. The shadow of the bars on the window threw a network over the yellow bathrobe. Many twinklings of eternity went by before she finally spoke. And even then the words were almost inaudible, more a movement of the lips than an utterance.

"Good-bye, Z."

9

The Cone

I didn't come out of the clouds until I was almost at the top of the Cone.

Although it was the middle of summer, Dark Mountain seemed buried in autumn. Down in the valley this was just an ordinary overcast day, probably muggy and humid, but here at an elevation of almost two thousand meters everything was clothed in a grayness that was less transparent than mist and somehow denser and more palpable. The sky literally touched the ground right here. The clouds were filled with minute drops, embryos of rain, that seemed to be moving in all directions, not just downward. If the temperature were to drop by just a few degrees, they would turn into crystals of snow. This actually happened now and then, though they always quickly reverted. During the summer on Dark Mountain you could go through all four seasons in one day.

In such weather it was not advisable to take long walks since you could easily lose your way. If they went out at all, people stayed close to the hotel, keeping to the asphalt paths where the lighting was on, even though it was just past noon. But I was not afraid of getting lost. I'd been coming to Dark Mountain for years, both summer and winter, and not a day would go by without a visit to the Cone. I was certain that I could find my way there even on a moonless night, though I'd never tried.

The Cone was a projection on the western slope, about two and a half kilometers from the hotel. The view from its peak was almost as fascinating as the one from the topmost craggy crest of Dark Mountain, accessible only to fully equipped mountain climbers. Owing to the Cone's almost perfect shape,

---

"The Cone." Written in 2000. Originally published in Serbian in 2000 as "Kupa" in *Nemogući susreti/Impossible Encounters*, Polaris, Belgrade, Serbia.

© Springer International Publishing AG, part of Springer Nature 2018
Z. Živković, *First Contact and Time Travel*, Science and Fiction,
https://doi.org/10.1007/978-3-319-90551-8_9

from which it derived its name, it seemed to be artificially planted there. As you approached, it didn't give the impression of being steep, but it was. The climb to the top thus required not only agility but considerable effort as well, even though the distance to be covered was less than one hundred and fifty meters.

These difficulties discouraged most of the hotel guests from visiting the Cone. On fine days they would walk to its foot, but only a rare few would decide to undertake the climb. In any case, the small, windy plateau at the top only had room for three or four people at most. When the weather was bad, like today's, I could count on having the Cone all to myself.

I came out of the cloud all of a sudden. I wasn't far from the top when it started to lighten. The grayness around me didn't thin or become more transparent, it just changed shade, turning a bright white. And then I suddenly rose above the foggy mass, squinting at the blinding radiance of the sun.

I stopped, still in cloud from the waist down, and waited for my eyes to adjust. Above me stretched the immeasurable, bright blue firmament, and as far as I could see below me was a motionless sea, its uniformity disturbed here and there by the islands of mountain peaks similar to the one I had just reached, forming a scattered archipelago in the sky. This panorama was worth all the trouble of the climb.

"Strange to find yourself above the clouds, isn't it?"

I started at the unexpected voice. I'd been so certain that I would be the only one at the top of the Cone that I hadn't even turned to look around, fixing my eyes on the horizon instead. The man was sitting on a rocky outcrop, his back turned to where I stood. It must have been the sound of my steps that told him I had joined him on the plateau. He was wearing a dark green jacket that blended in with the color of the surrounding grass and low bushes. His hair was gray and longish, partially covering his ears.

"It isn't usually crowded above the clouds," I replied, making little effort to hide my displeasure. I wasn't pleased at having to share the Cone with someone just then. I sat down on a patch of grass behind the stranger, feeling beforehand to see if it was wet. Among the thick tangle I found an empty can of soda pop carelessly left there. I picked it up and threw it into the depths below. I was aware that this was just as careless, but it seemed somehow more fitting for garbage to be found anywhere but here.

"No, it isn't. I liked it best when I could be alone here, too." He said this without any reproach in his voice, which made me feel awkward. In fact, he could consider me the intruder since he had reached the top of the Cone first. "But I won't bother you for long. I'll be leaving soon."

"You don't have to go because of me," I said obligingly. "There's room for both of us."

The man did not reply, so we fell silent, gazing into the distance. The warmth I started to feel wasn't just from the strenuous climb. It was considerably warmer here in the sun than down in the clouds. I did not unbutton my jacket, however, even though I could feel the sweat breaking out; the wind that never seemed to die down here at the top might blow through me.

"I haven't been on the Cone for a long time," said the man pensively, as though addressing someone invisible in front of him, rather than myself. "The last time I climbed up here I was your age."

I stared at his back in amazement. How could he know my age when he hadn't turned around to look at me? Probably by my voice. I hadn't seen his face, either, but even without the gray hair I could easily tell by his hoarse, wheezing voice that he was well into his sixties.

"You've missed quite a bit," I said with a smile.

"I know. I'm trying to make up for it now. I'm visiting places that meant something to me in the past."

"Did you stay at Dark Mountain very often?"

"Yes, at least twice a year. I never did learn to ski, although I loved to take long walks."

"Me, too. I'm not the least bit bothered by not being able to ski. Walking is just as pleasant, and you need a lot less equipment."

The gray head nodded in front of me. "At first I went for walks in different directions. But after I discovered the Cone, I gave up all the other places. I started coming here every day, almost like a ritual. Over time it became a real obsession. The only thing that could stop me was a snowstorm."

Strange, I thought. It's as if the old man was describing my own experience. I never imagined I'd ever find such a kindred spirit. Most people think I'm an oddball because of my pilgrimages to the Cone. There was, however, one important difference.

"But it seems you got over your obsession. If I understood correctly, you stopped visiting the Cone. What prevented you from coming?"

The man did not reply at once. When he finally spoke again, his voice became softer, so that I had trouble making it out against the howling of the wind.

"I experienced something unusual here. Afterwards there was no sense in coming here any more."

I expected him to continue, but as the old man didn't elaborate, I had to curb my curiosity. For some reason he clearly did not want to talk about it, and good manners would not let me probe. We passed another few minutes in silence. I could feel the skin on my face start to prickle under the strong mountain sun. I should have brought some sun screen, although I hadn't actually expected the top of the Cone to be above the clouds.

"I like to return to places that mean something to me, too," I said at length, just to keep the conversation going. Although he had said he would be leaving soon, the old man continued to sit there, and it seemed silly not to talk while we shared this cramped space. "But it's never like it was the first time. The place might be the same, but the time is always different. That can't be helped, I'm afraid."

"Except if you return to some place at the original time," he said, his voice still low.

"In the past?" I asked with an inadvertent cry of disbelief.

The old man raised the collar of his jacket a little to protect himself from the strong wind that had just come up. Although quite blistering, the sun was deceptive. It would be easy to catch cold.

"Yes, in the past."

"Then it really would be just like the first time. Except it isn't possible. You can't go back into the past."

"Even so, if you were offered the chance to go back, which time in your life would you choose?"

My eyes began to skim over the endless landscape that surrounded me. Far to the east the sun had finally triumphed over the clouds and now wooded hills could be seen though the mist. By late afternoon it would clear up here, too, and Dark Mountain would return to summertime.

"I've never thought about that," I said. "I don't know, maybe some point in my childhood. I would probably like to see myself as a boy." I stopped for a moment, staring blankly at the gray shroud beneath me. "That would certainly be strange—to meet your own self."

The old man turned his head a bit towards me, enough so that I could see his thick gray beard and sunglasses, but then he faced forward again.

"Why your childhood? Do you feel you were happier then than later in life?"

"It's hard to say," I replied after a brief hesitation. "Perhaps more innocent. There were happy moments later on, of course, but they lacked that early innocence. It seems to be more and more precious as time goes by. But what about you? Which time in your life would you go back to?"

The man shrugged his shoulders. "At my age childhood is already far away and faded. I think I would choose something closer, something I remember better. I was very happy when I came here to the Cone. Perhaps even innocent, in the sense in which you talk about your childhood, although it didn't seem like that at the time. In any case, I left innocence behind me forever on the Cone. I would be happy to meet myself again from that time."

I wiped the sweat off my brow with the back of my hand. "I bet the other one would be just as happy. Maybe even more so. It would be a very useful encounter for him. You could tell him first hand what awaits him in the future, what he should stay away from, what he should avoid."

"Oh, no, not at all," replied the man quickly, raising his voice a little. "I wouldn't tell him that at all."

"You wouldn't tell your own self what the future holds in store?"

"No."

"Why?"

"Because I would ruin my own life if I did. The encounter itself would be extremely risky. It would be best if he didn't realize who he'd met."

"I don't understand."

"If I told him what the future holds, I would be depriving him of the foundations that make life possible. Everything would become preordained for him, inevitable. He would lose not only hope but fear. And how can you live without hope or fear?"

"But what if, for example, there was some great misfortune or suffering awaiting him, that could easily be avoided if he was forewarned? Would you allow that to happen?"

"Of course."

"Wouldn't that be cruelty towards your own self?"

"Perhaps. But there is actually no choice. You cannot prevent what has already happened, can you?"

I didn't know what to reply. I had the vague feeling that there was some sort of paradox involved, but I couldn't put my finger on it. No doubt it all hung from the unfeasibility of the initial assumption about returning to the past.

The old man stood up and so did I. He was approximately my height, perhaps a bit stooped owing to the weight of his years. He picked up something he had been sitting on, and as he brushed off the bits of grass I realized it was a book. Before he put it in his pocket, I managed to read the large title—*Impossible Encounters*—but not the name of the writer.

He stayed a few moments more, staring at the sea of clouds that had now gently started to stir and thin out. Then he turned towards me and we were face to face for the first time.

I couldn't really see much of his face. It was hidden by his beard and the large sunglasses. Only his forehead was uncovered—it was even higher than mine because the gray strands had receded quite a bit towards the crown of his head.

"It's time to leave," he said. It might have been my imagination, but his voice seemed to tremble slightly, just like mine on the rare occasions when I am excited. He extended his hand and I took it in mine—a slim, bony hand, just like mine will probably be when I reach his age. "The Cone is all yours. Enjoy it while you can. One never knows what the future will bring."

"I'm glad we met," I said, more softly than I intended.

"I'm glad, too. Very glad."

He let go of my hand with some hesitation, almost unwillingly. Then he turned and headed down the steep slope, without looking back. He walked slowly, carefully. Like an old man. When he disappeared into the cloud, I felt a sudden lump in my throat.

I stayed on the Cone for a long time that day. Almost until dusk. By the middle of the afternoon everything below me had cleared up. I slowly absorbed the endless, luminous panorama surrounding me. I wanted to remember it well. I intended, of course, to come again next day, but the old man was right: I did not know what lay in store for me. What if something prevented me from coming? What if a long time, several decades, passed before I happened to climb the Cone again?

All fiction translated from the Serbian by Alice Copple-Tošić.

10

Annotations 2

At the very beginning of my creative writing course at the Faculty of Philology, University of Belgrade, I used to ask my students a seemingly simple question: "Why do we write prose?" Every time, there was a variety of imaginative attempts to answer, but only rarely did they come near to the truth—that we write prose because we still haven't invented any better way to provide answers to the most profound existential questions. No other discipline, artistic or non-artistic, regardless of how illuminating it might be in this respect, can really compete with the art of prose.

There is a very special kind of literary gnoseology that provides us with a deep understanding of the complex and complicated dilemmas, ambiguities, and paradoxes of human existence; much deeper in fact than those offered by philosophy, religion, or the sciences. In the entire art of prose there are basically only two main themes—love and death—around which everything else circles. Is there any connection between our ability to love and our mortality? Is the former a kind of compensation for the latter? Would we still be able to love if we were immortal? You don't have to be a philosopher, a theologist, or a scientist to cope with these questions. Just read any great work of literature. It is only there that you may hope to find truly good answers.

The art of prose is no less useful in handling ideas which apparently belong to the domain of speculative thinking. Time travel, for example, or first contact. In the 20th century, science fiction had its say about these two themes. It was mostly imaginative and original, but certainly not all there was to say due to SF's generic limitations. To avoid them, a new, non-generic "fantastika"[1] was required, which started to take shape in the early years of the 21st century.

Being its humble practitioner, I had an opportunity to try to make the most of the literary potential of the time travel and first contact themes. My mosaic novel *Time Gifts* and short story "The Cone" are not about how to achieve chronomotion (that would be SF), but whether it has any meaning. A human meaning. Do those who receive various time gifts from a mysterious visitor really benefit from them? Are they now any happier? Should an older version of the protagonist disclose to his younger self what the future has in store for him? Or is ignorance of the future the only thing that makes life possible?

Similarly, is there an aspect of first contact that has been overlooked or neglected so far? We have approached the problem of the Grand Silence of the Universe from all imaginable angles—save one. In my stories "The Puzzle" and "The Bookshop" I used art as the key to the solution. Is a series of drawings inspired by music the long-awaited signal from the stars that has finally arrived? Is a literary work, barely written, able to provide a shortcut to another cosmic island inhabited by Others?

Difficult questions that can be properly asked and answered solely by the noble art of literature. It is only right that it should have the last word.

---

[1] I decided to retain here the original Serbian term "fantastika" because apparently there isn't a fully satisfactory English equivalent. The "fantastika" encompasses all non-mimetic types of narratives, in the sense that worlds imagined in the literary works of this sort don't fully coincide with what is generally considered to be reality. According to certain statistical studies of literature, nearly 70 percent of everything that has ever been written belongs to one or another type of "fantastika."